READINGS IN AGRICULTURE

Linda Houston

The Ohio State University
Agricultural Technical Institute

KENDALL/HUNT PUBLISHING COMPANY
4050 Westmark Drive   Dubuque, Iowa 52002

Contents

PREFACE

Readings in Agriculture is a collection of nonfiction articles in many areas of agriculture from a variety of current periodicals related to the technical aspects as well as the business facets of the profession. These readings are intended for use in writing courses in agricultural schools or departments as well as content-area courses dealing with agricultural topics. Summarizing or analyzing the articles, researching the topics in the articles, and using the articles as a starting point for developing other issues which relate to agriculture can develop the students' ability to understand the process of writing and critical thinking as a means of inquiry.

These readings are grouped loosely into five categories; they go beyond the "how-to" involved in the technical aspects of the field and often may overlap with each other, thus highlighting the interdisciplinary nature of the publication.

There are some sample exercises in the Appendix of the book given only as examples of ways to engage students in analyzing the readings. Giving students a variety of ways to respond helps them to read critically. Classroom discussions, group responses, panel presentations, debates, and further research on the topics introduced in the articles are other techniques which can be utilized with this reader. The publication is not meant to be prescriptive; instead, each instructor is encouraged to use the readings for student-centered learning, adapting the book to the needs of the individual student and to the course for which the text is used.

Acknowledgments

I am grateful for the cooperation of the publishers and authors that gave us permission to reprint their articles. They have thus furnished the students with a wide spectrum of reading experiences and thus an awareness of potential sources of information.

A special thank you goes to Seth DeHoff, a graduate of The Ohio State University Agricultural Technical Institute and a current student at The Ohio State University in Columbus, OH, for his commitment in helping select the articles which appear in this publication. Because of his ability to choose articles which would be of interest to students studying agriculture, this collection represents readings of interest to students. Without his help, it would have been difficult to complete this project.

Finally, I am grateful, as always, to my students who over the years have helped me see the need for a reader dedicated to their interests in the agricultural profession—a reader that can be used in their writing course as well as other content area courses.

PART I

PLANT INDUSTRIES

Analyze This

David Marquardt

Back in Soil Science 101, the attitude toward fertilizer was "Put it on; don't put it off!" Yet, the idea that something is better than nothing is not fertility management. Growers and landscapers need to treat the soil system as if it were a three-legged stool: If the soil's chemical, physical and biological properties are not balanced, the stool will collapse. Treating all soils as if they were inert has resulted in chlorotic and stunted plants, as well as landscapes lacking vigor and color.

Trade journals are excellent for providing the "how to" of estimating the amount of fertilizer to use, but they cannot replace conducting proper soil tests and using fertilizers to balance nutrient deficiencies. The same holds true for testing composted organic soil amendments, such as yard waste and spent mushroom composts. These amendments have their own chemistries. No professional would recommend applying 10 pounds per 1,000 square feet of an unknown fertilizer. So why should professionals apply 1 inch of an organic compost as a soil amendment when they have no idea about its chemistry?

As the owner of Bio-Grow Technologies, a horticulture consulting company in Hinckley, IL, I evaluate the soil-testing and fertility-management practices of nurseries and landscape companies in Chicago and northern Illinois. I continually find sites with poor plant performance, and it's often the same scenario: A landscape has heavy, high-alkaline clay soil compacted during construction. The soil is amended with organic compost, sand and sometimes pea gravel. While this seems to be the perfect soil environment, the plants are not healthy. What went wrong?

It turns out that an audit of the soil chemistry has not been performed, so the professional does not know the native soil's fertility. In addition, the amendment has not been tested. Without testing soils and amendments, continued fertility shortages or, even worse, excesses will develop.

For example, the professional often does not address the ratio of carbon to nitrogen in hardwood mulches. As a noncomposted bark and wood product, hardwood mulch is high in carbon, the breakdown of which requires nitrogen-consuming bacteria. If nitrogen is not provided in the soil mix, these bacteria will begin to feed on the plants themselves, causing them to become yellow and stunted.

The key is to use the appropriate amendment for each planting environment. Professionals know they can't plant a flowering dogwood in just any soil, so why do they continue specifying the same amendments for every soil? I see contractors

---

COMPOST AUDIT RESULTS*				
Test Performance	Yard Waste Compost 1	Yard Waste Compost 2	Reference	Mushroom Compost
Cation exchange capacity	40.58	34.06		36.86
pH	7.8	7.4	6.3	7.4
Percent of organic matter	28.07%	26.9%		57.2%
Anions				
Soluble sulfur	135 ppm	122 ppm	35 ppm	1,120 ppm
Easily extractable phosphorus	637 ppm	348 ppm	35 ppm	450 ppm
Percent of Base Saturation				
Calcium	49.85%	57.97%	68%	60.9%
Magnesium	23.18%	18.3%	12%	14.69%
Potassium	22%	17.7%	5%	16.05%
Sodium	1.37%	2.07%	0.5%	4.35%
Other bases	3.6%	4%	4.5%	4%
Hydrogen	0%	0%	10%	0%
Extractable Minor Elements				
Boron	3.57 ppm	2.98 ppm	1 ppm to 2 ppm	1.51 ppm
Iron	241 ppm	175 ppm	225 ppm	98 ppm
Manganese	70 ppm	117 ppm	60 ppm	165 ppm
Copper	3.8 ppm	2.14 ppm	5 ppm	2.25 ppm
Zinc	17.96 ppm	31.7 ppm	10 ppm	16.3 ppm
Soluble salts	3.13 ppm		1 ppm	7.7 ppm
Nitrate-N	126 ppm		10 ppm	
Ammonium-N	132 ppm			

*The values in the Reference column that are not constants have been left blank. Other blank areas spending tens of thousands of dollars correcting fertilizer problems on commercial landscapes. This comes after years of plant replacements and frustration. Corrective steps should begin when specifying design and installation. Specifying fertilizers and organic amendments without properly matching them to the soil's physical and chemical needs yields poor performance in the landscape.

Furthermore, soil and amendment audits are necessary because compost is often unlabeled. Thus, green industry professionals need to enlist the help of industry consultants, as well as send samples of their soil and compost to a soil analysis laboratory. Many labs are available throughout the country. Such services are as important to one's business as a shovel and rake. When choosing a laboratory, an important rule of thumb is to choose one not connected in any way to a particular product. Independent laboratories are the best choice.

Once the soil and amendments are tested, professionals can begin the matching game. Remember, the basic rule of physics holds true: For every action, there is an equal and opposite reaction.

Match Game

To illustrate my point, I am going to discuss the results of tests conducted on samples of two yard waste composts and a mushroom compost sample. Bio-Grow Technologies collected the samples in Missouri and Illinois, and an independent laboratory conducted the audits. Results are shown in the table above.

The table's Reference column indicates standard values, as determined by independent consultants. The standard values in this table are derived for woody ornamental sites and should be viewed as a point of reference. Ideally, you should match the soil and amendment audits to achieve results as close to the reference as possible.

The tests reveal that although the yard waste samples came from about 400 miles apart, the samples have similar values. Thus, similar products that undergo comparable compost procedures generally have similar resulting chemistries.

Optimum nutritional release occurs when the soil pH is 6.2 to 6.5. As the pH rises, iron and manganese become increasingly unavailable. Iron becoming insoluble can lead to chlorosis, and in-sufficient manganese limits a plant's ability to produce chlorophyll during photosynthesis. The table shows that both yard waste samples are alkaline, with pHs of 7.4 and 7.8. If one of these amendments was added to a typically alkaline soil, let's say with a pH from 7.2 to 7.6, the soil would become highly alkaline and the availability of micronutrients would be severely limited. To reduce the pH, some professionals mistakenly overapply sulfur. What is not realized, however, is that pH is a function of the balance of the earth's alkaline metals rather than a lack of sulfur—and excess sulfur is detrimental to plants.

The same holds true for excess phosphorus. When professionals use amendments with excess phosphorus, micronutrients may again become unavailable—especially iron, manganese and zinc. The result is chlorosis. This condition is difficult to overcome because it may take years for plants to use up the phosphorus or for it to be broken down to acceptable levels. Adding acidifying amendments such as iron sulfate to alleviate chlorosis only provides a short-term solution. Adding calcium may bind some phosphorus and thereby free up some iron. But this, too, only offers short-term results and is costly in terms of materials and labor. The table shows that the easily extractable phosphorus levels in both yard waste samples are far above that of the standard. Thus, misuse of either of these composts will likely adversely affect plants.

The yard waste composts also contain excess potassium and boron. High potassium levels limit the soil's ability to hold calcium and magnesium. Additionally, recent research has shown that when potassium and boron are excessive, the plants show yellowing of leaf tips, necrosis, marginal and interveinal leaf scorch, and premature leaf drop. Irrigation water that contains high levels of boron only amplifies the problem. According to the compost audits, boron is already excessive in both yard wastes, and more frequent applications of nitrogen, phosphorus and potassium will not correct the situation. In the nursery fields, excess potassium causes lack of caliper, and lack of copper leads to trunk splitting. A sod and amendment audit is invaluable in preventing such problems.

Soluble salts are another factor to consider when choosing organic amendments. Excess soluble salts injure plants. The table shows that one of the yard waste samples had an excessive soluble salt level of 3.13 parts per million (ppm), compared with the standard reference of 1 ppm.

Container plants can be leached with clear water to remove excess salts. Yet, leaching is not possible in the tight, compacted soils of a landscape. Salts in these soils move down only as far as the bottom of the prepared bed, where they accumulate and cause root damage, resulting in minimum nutrient uptake. This results in slow root development, reduced vigor, stunted growth and midday wilting. Landscape contractors typically try to increase plant vigor by applying fertilizer, but this often stresses the plant and reduces its ability to feed itself.

Let's now look at the audit of the mushroom compost sample, which illustrates that this amendment must also be used with caution. As illustrated in the table, not only does this sample have excess phosphorus and potassium, it also has excess sodium—perhaps the hardest problem to correct. Excess sodium takes the space of calcium, magnesium and potassium, resulting in a higher pH. As mentioned earlier, an alkaline pH limits the availability of many nutrients. Removing sodium requires the use of calcium, magnesium and potassium, as well as a solublizer such as sulfur. The most common remedy is gypsum, which is calcium sulfate. This process takes a great deal of water and time and is often complicated by slow percolation rates.

Understand what is being said here. These products aren't as much bad soil amendments, as they are misused in this situation. In areas where soils are well-drained and organic matter levels are low, these products may be very valuable. Many low cation exchange capacity, well-drained soils fail to maintain fertility. Products inherently high in organics and fertility and used at appropriate levels should create a good planting environment. The key is to know your planting environment and match it to your amendments.

Nursery and landscape professionals should take a proactive approach to fertility management. Everything is interconnected: Micronutrients, cation exchange capacity, level of organic matter, the type of plants and the soil percolation rate are just some factors that affect soil chemistry. Effective fertility management can boost the success of a business, and it is far more cost-effective than having an inadequate management system in place. Do not get caught up the philosophy of some professionals: "We never have time to do it right, but we always have time to do it twice."

Landscape and Nursery Dialog

The GMO Debate: Where Do We Stand?

Pablo Jourdan

Perhaps one of the most controversial and contentious issues in Agriculture in the past year has been the role Genetically Modified Organisms (GMOs) play in current crop production practices. While the relevance of this topic to grain crop production and food science may be obvious, does the controversy have any significance for the green industry? How much have you been following this topic? A quick (and admittedly somewhat superficial) scan of trade publications, news reports, and websites revealed little coverage of the topic in the green/ornamentals industry. Yet a question posed by some (for example, David Kuak of GMPro magazine) is very relevant: Were a genetically modified ornamental crop available now, would you grow it or even use it in a landscape? Perhaps with our concerns centered on labor shortages, regulatory compliance, the drought, etc., we have not had time to think much about this topic. Nevertheless, the GMO debate/controversy is significant and its ultimate outcome can have an impact on our industry. This article is the first of what I hope can be a series of dialogs on the topic of GMOs, biotechnology, genetic engineering (or whatever one chooses to call it) within the context of the green industry. As before, we encourage comments and feedback to what you read here at the LAND website (http://www.hcs.ohio-state.edu/land).

A rather common refrain in the debate, voiced particularly by scientists and leaders of the biotechnology industry is the large gap in knowledge about the topic between the professionals and the general public. While "biotechnology" has been talked about for over 20 years now, most individuals not directly involved with it find much of the jargon impenetrable and some of the technology's potential exciting, daunting, or frightening, depending on personal philosophy. "Ambivalence" could have been used at one time to describe the feelings of many toward biotechnology, but that hardly applies now. It is unarguable that for the debate to be fruitful, participants should have a basic understanding of the topic. To that end, this article highlights key concepts and terminology; later articles will address specific applications, the social/political context, and environmental concerns.

It is unfortunate, in my view, that the term "genetically modified organism" has become the common moniker for the topic because I could argue that the origin of human civilization and its roots in agriculture/horticulture traces to our ability to genetically modify plants and animals. The entire process of plant and animal domestication has been primarily one of genetic selection: we chose what individuals to keep for the next generation. Gardeners and horticulturists have a long history of "modifying" plants and even "creating" entirely new ones. Some examples:

- ❏ The Red Horsechestnut, *Aesculus x carnea* is the product of two species that occur naturally in two different continents; one, the Red Buckeye, *Aesculus pavia,* is native to North America; the other, the European Horsechestnut, *Aesculus hippocastanum,* is native to Europe. While its origin is undocumented, some gardener, probably in Germany, grew the Red Buckeye and the Horsechestnut in proximity to each other and the result of this "unnatural liaison" (only because these plants were brought together by people, not by natural events) is one of the most popular trees in Europe.
- ❏ The Saucer magnolia, *Magnolia x soulangiana,* was derived from a cross between *Magnolia denudata* and *Magnolia liliiflora* made in France in the early 1800's.
- ❏ The Leyland Cypress, *XCupressocyparis leylandii* is the product of a cross that occurred in Wales in 1888 between *Cupressus macrocarpa* and *Chamaecyparis nootkatensis.*

None of these plants would have likely existed were it not for the deliberate actions of people. This list could go on and on to include herbaceous ornamentals such as the geraniums; and if we were to expand into food crops, we'd have to mention the very existence of corn and wheat as possible products of genetic modification by humans.

So genetic modification of organisms has been an ongoing process for millenia, but the tools for modification expanded most dramatically in the 20th century resulting in a much more precise and efficient process. These tools are:

- ❏ the application of genetic principles (first elucidated by Gregor Mendel in the 1860's) in the practice of plant breeding,
- ❏ the subsequent detailed analysis of genetic elements that resulted in the discovery of DNA, the chemical "blueprint" for life (early 1950s) and the more detailed understanding of genes (ongoing),
- ❏ the development of plant tissue culture techniques which permit the culture of individual cells and their controlled development into whole plants (1930's to 1960's)
- ❏ the discovery that a common plant disease (crown gall) caused by a soil bacterium, *Agrobacterium tumefaciens,* provided a mechanism to transfer genes from one organism—the bacterium—to another—the host plant (early 1980's).

With additional refinements in technology, it has been possible for some time now to introduce into a plant any gene of interest, regardless of the origin of that gene (be it from a virus, bacterium, fungus, or animal—including humans). So GMOs might be defined as "plants or animals which have had their genes changed in the laboratory by scientists."[1] Modification of genes in an organism leads to modified characteristics such as flower color, height, disease resistance, herbicide tolerance, etc.

Until the advent of this technology, most genetic manipulation was limited to closely related species that could be crossbred, as indicated in the above examples. This process is very unspecific and along with traits of interest could come potentially undesirable characteristics that often takes years of additional breeding to eliminate. This approach has been amenable primarily to annuals or some herbaceous perennials that have relatively short life cycles, but it has had a more limited application to long-lived plants such as trees and shrubs. One of the key attributes of genetic engineering that has made it so compelling is precision and the accompanying increased efficiency in modifying the genetic characteristics of plants. A single gene influencing a single characteristic can now be manipulated within the plant. However, since the technology is still in its infancy, the attributes of precision and efficiency must be considered relative as we have much yet to learn about how to effectively manipulate most genes. A typical plant, whether a small weed similar to Sheperd's Purse or a large tree like a Burr Oak, contains about 100,000 genes that interact in complex ways with each other and with the environment to define the plant and its characteristics.

In summary, the terms plant biotechnology, plant genetic engineering, genetically modified organisms, refer to a set of tools that extend our ability to manipulate plants. The genetic modification process might follow this general pattern:

❑ Genes are manipulated in a laboratory by changing the coding sequence, altering its architecture and therefore affecting its ultimate function in the plant. Our current understanding of gene function together with an impressive battery of molecular tools allow us to modify gene(s) by controlled biochemical reactions, and once introduced into plants, direct the gene(s) to be functional either throughout the plant or only in the leaves or roots or flowers or fruit. It is also possible to modify the relative activity of an introduced gene or of a gene already in the plant; the gene can be made much more or much less active, even to the point to shutting it down. A common practice is to couple a desired gene, for example, one that makes plants dwarf, with another gene that inactivates an antibiotic; the latter is needed in the tissue culture step of the process (read on).

❑ The modified gene(s) can be introduced into plants by adapting the natural *Agrobacterium* (crown gall) infection process or through a "biolistic" process where DNA can be coated onto microscopic "bullets" that are then shot into plant cells in sterile culture with a "gun." The general principle is the same in both systems: a carrier (the bacterium or a "bullet") introduces the gene into a plant cell and the gene finds its way into the appropriate internal cell machinery needed to function.

Because the transfer process is rather coarse, only a few among thousands or millions of cells receive the new gene. Here's where the coupled gene that inactivates an antibiotic comes into play: to select for the few cells that were "transformed," that is, they received the new gene, the cells are cultured in the presence of the antibiotic which kills all that do not have the introduced antibiotic resistance gene. The gene we actually want is selected along with the antibiotic resistance.

❑ The next step is to stimulate the surviving cells to become whole plants. For many plants it is possible to maintain loose cells in sterile culture and then stimulate the cells to "reorganize" into whole plants using techniques similar to those for micropropagation. The resulting—"regenerated"—plants should still contain the gene of interest (as well as the antibiotic resistance gene).

❑ A series of tests are then used to confirm that the plant carries the desired gene and it is functioning properly. The plant can now be propagated by the most appropriate technique and eventually introduced into the market.

The tools described above have opened up opportunities for altering plant characteristics that were unimaginable just a couple of decades back, but they have also led to great concern about the potential risks that the technology poses. Given humans' somewhat chequered history in harnessing powerful technologies (for example, nuclear power), it is fair and appropriate that concerns and risks be thoroughly discussed and assessed. Both the technology and the issues are complex. I'll attempt in future articles to address other aspects of the tools (such as what traits are ready for manipulation, what are the bottlenecks) and their application (such as safety and environmental impact).

❑ Notes

1. From the BBC on-line FAQ about genetically modified foods (http://www.bbc.co.uk).

Managing the Feature Tree

John Ball

I remember tagging along on an appointment with one of our consulting arborists. We had been examining a declining oak and were running a bit late, so he asked if I minded riding along to his next appointment before going back to the office. The call was for a removal quote. "Shouldn't take long," he said.

When we arrived at the house, it was readily apparent which tree was being removed. In the middle of the front yard stood a dying American elm (Ulmus *americana)*. The owner met us as we pulled up and said all he wanted was our "best price" to remove the tree and haul away the brush. It did not look like a difficult task. The tree could be felled rather than pieced, and there was a lot of open space in which to work. No power lines to work around, no buildings, fences, or even sidewalks to worry about. The arborist made a great presentation about what our company could do: a safe, efficient removal and a thorough cleanup—for $480. But the owner was not interested; someone else had stopped by earlier in the day. This person had noticed the big yellow "X" on the tree—the mark used by the city to identify trees condemned due to Dutch elm disease—and figured he could use the wood for his fireplace. He would remove the tree and haul the brush out to his place for $100. We did not get the job.

That's the trouble with bidding removals. In the eyes of the tree owner, the end product of a removal—a stump with a minimal amount of debris and divots—is the same regardless of who is hired. The issue of who to hire is not based on skill level with one obvious exception: large trees that overhang the house. It is obvious to almost everyone that these trees require expertise. Otherwise, in the removal business—price rules.

Unfortunately, much of the tree care industry is focused not on care but on cutting. In a recent survey of upper midwestern tree companies in the United States, as a percentage of sales volumes, removals were number one (Ball, unpublished data). Even on the national level, removals rank number two in sales (NAA 1988). Removals do not lend themselves to repeat business. After all, with the possible exception of white poplar *(Populus alba)* and a few other species, you can remove a tree only once if the stump is ground. True, we can, and should, take the opportunity with removals to explain that proper care may prolong the lives of a client's other trees, and removals often can be followed by plantings. But removals are usually a dead end for that client relationship. This situation is just the opposite in the lawn care industry, for which

70 percent retention of clients from year to year is expected (Clark 1999). We lag behind the lawn care industry in other ways as well. As an industry, we capture only about 15 percent of the dollars spent on residential landscape maintenance. And the average household that employs landscape maintenance services, either tree care or lawn care, spends more on their lawn, $533 per year (Clark 1999), than they do on their trees, $434 (Hogan 1999).

Much of the tree care, as opposed to tree removal, industry is fueled by two important entities: clients able and willing to pay for the services and trees the clients believe are worth the expense. This segment of the tree care market includes what we have come to call the feature tree.

What is a feature tree? Webster's dictionary begins its definition of feature as "the structure, form, or appearance" and continues with "a prominent part or characteristic." Thus, a feature tree is best described as one whose appearance makes it a prominent part of the landscape. The tree may be exceptionally large—a remnant eastern white pine (*Pinus strobus*) towering 120 feet above the home, or it may be old—a white oak (*Quercus alba)* more than 400 years old. The feature tree may also have historical value: a tree planted at the founding of the community or perhaps one that is a local landmark. But most feature trees do not fit any of these categories. They are neither exceptionally large nor old, or even historical. Their prominence is due solely to their appearance, their beauty What defines beauty in a tree? What people find attractive in a tree has been studied, and feature trees tend to have certain physical characteristics in common regardless of species.

What do people like about trees? The most important attribute for a tree is appearance—it is pleasing to look at (Sommer et al. 1990). This is not surprising: After all, these trees are called *ornamental* for a reason. What do people find pleasing about trees? Their shape. When viewing different shapes of trees (from columnar to globe), people usually state that the most appealing shape is a spreading one, a tree that is broader than it is tall. In addition, an attractive tree branches well below half its height, and the branching pattern is layered (Summit and Sommer 1999). This shape describes the appearance of two of our most cherished trees, the beeches (*Fagus* spp.) of the U.S. northeast and the live oak (*Quercus virginiana*) of the south. Not too surprisingly, the most appealing tree shape is used in the Society of Commercial Arboriculture logo. If you're looking for a tree to put in your advertisement or logo, you could not go wrong with one that is similar. It is the shape that people instinctively like.

Regardless of the reason for its special value, the feature tree is one whose care is important. Removal is not an option. The loss of such a tree to its owners is something that gives appraisers fits when attempting to apply the trunk formula. Pricing these trees by the square inch would be the equivalent of appraising the *Mona Lisa* based on the square inches of each color of paint. Although we can put a monetary value on such a tree or painting, they are priceless and irreplaceable in terms of beauty and value to society.

The feature tree is usually a mature specimen, owing to the importance we place on size and longevity. Because such trees are mature, our care is focused on maintenance rather than growth. We need to be in it for the long term when entrusted with trees that have the potential to outlive their owners and us. With the feature tree, our value as arborists is based not on the amount of brush we create but on the knowledge we hold.

Aboveground care of the feature tree is primarily focused on three practices: pruning, support systems, and, in some regions, lightning protection. The objectives of pruning mature trees are quite different from those of pruning young trees and are much more limited. The mature tree has reached a delicate balance among all its parts—canopy, trunk, and roots—a balance that should not be disturbed without much forethought. When pruning a mature tree, focus on the four Ds: dead, defective, dying, and diseased. This focus encompasses what has come to be known as "crown cleaning."

Thinning, the selective removal of branches to increase light penetration (ISA 1995), must be cautiously applied on mature or overmature trees. Pruning is more than the indiscriminate removal of branches. When I ask arborists why they are thinning a mature sugar maple (*Acer saccharum*), beech, or similar tree, the common response is, "We need to get more light in the tree." When further pressed, they may add, "Those shaded branches are parasitic—taking more food than they are making—more light is what is needed."

But tree species differ in their light requirements. Sugar maple leaves can operate efficiently at lower light intensities than cottonwood (*Populus* spp), for example. Thinning a sugar maple crown to allow increased light penetration may be counterproductive. The interior leaves are adapted to the lower light intensity and opening the canopy may not result in an increase in carbohydrate production. Regardless, these interior branches are not parasitic. The question has long been asked if lower, interior branches contribute carbohydrates to the rest of the tree, parasitize carbohydrates, or are neutral (neither helping contribute nor taking). Pruning studies have generally found that the overall effect is neutral—the removal of these branches does not increase growth of the overall tree (Sprugel et al. 1991).

Leaves, and their associated twigs and branches, survive as long as they can produce at least as much carbohydrates as they use. Thus, branches generally can be described as exporters, not importers, of carbohydrates. If the foliage on a heavily shaded branch cannot produce enough carbohydrates to support it, the branch dies and is shed. It cannot extend its life by robbing carbohydrates from other, more sunlit, branches.

The definition of crown cleaning also includes the removal of these "low-vitality" branches. However, I prefer to call these branches *nonfunctional* rather than *low-vitality*, although both terms can apply. Functional is defined as contributing to the greater whole. Obviously, nonfunctional branches no longer contribute to the greater whole—the supporting trunk and roots. Their removal can be beneficial because they are not exporting carbohydrates to the trunk and roots, and they may provide a site for pathogens or insects to exploit.

Support systems can play a major role in maintaining mature trees. Mature trees often acquire numerous structural defects over their lifetimes. Sometimes corrective measures may call for the removal of a branch or codominant stem, but perhaps such actions would remove so much foliage they would jeopardize the tree's ability to adequately maintain its other parts. In these cases, a support system may be a better option. Unfortunately, many tree care companies shy away from such work. Cabling and bracing requires specialized arboricultural knowledge, skills, and equipment. It is also expensive and creates certain obligations and liabilities for the arborist. Cabling and bracing is not appropriate for all trees. Nevertheless, the proper application of support systems can extend the life of many mature trees. The same holds true for lightning protection. I worked in one community in which a major threat to mature white pines was having the top of the canopy blasted apart by lightning.

The belowground tree demands as much attention as the canopy. Feature trees often are isolated in a grass-dominated landscape, which can create intense competition between these two vegetation types, as well as conflict in the client's landscape objectives—a lush green lawn and trees. Sometimes, it is hard to manage for the optimal performance of both. Once, when asked about the impact of radial trenching on the appearance of the lawn, I pointed out to the clients that I could replace their entire lawn in three weeks or a weekend if I used sod, but it would take more than a lifetime to replace their mature oak. Where should the emphasis be placed? Faced with this logic, many opt in favor of the tree.

Belowground management of trees is relatively new territory for many tree care companies. Currently, fertilizing is a common practice and, in many companies, perhaps the only belowground practice. This is unfortunate because there has been a tremendous expansion in knowledge about how and where roots grow. Modifying the site, such as improving drainage and aeration through the use of vertical mulching and radial trenching, is increasingly studied, and the benefits and limitations of these practices are better understood (Kalisz et al. 1994; Watson et al. 1996). The benefits of mulching as a means of modifying the soil environment are now better known (Greenly and Rakow 1995). Mycorrhizal inoculations are also showing promise as a means of improving the health of mature trees (Smiley et al. 1997). While much work remains to be done regarding the application and benefits of these treatments and practices, they do offer new opportunities for arborists in their care of trees. In the next century, perhaps our care of the belowground tree may become as common as our present-day care of the aboveground tree. Hopefully this belowground care will be applied as part of a holistic Plant Health Care program—because treatment to one part of the tree, be it above or below ground, can affect the rest of the tree.

Finally, the management of feature trees is based on the needs of the client. After all, the designation *feature tree* is a social term, not a biological one. In caring for feature trees, arborists can be valued for their knowledge as well as their skills. I remind arborists that consumers tend to value or pay more for services that they don't know how to do themselves and that they believe require special knowledge, and less for those services they merely do not want to do. Extending the lives of such trees requires us to be at the cutting edge of arboricultural knowledge and skills. The feature tree can bring out the best in us, as arborists, and can allow us to demonstrate the value of our profession.

❏ Literature Cited

Clark, K. 1999. Lawn and order. *U.S. News & World Report* 126(17):58–60.

Greenly, K., and D.A. Rakow 1995. The effect of wood mulch type and depth on weed and tree growth and certain soil parameters. *Journal of Arboriculture* 21:225–232.

Hogan, G.K. 1999. Tree and tree-service sales. *Grounds Maintenance* 34(3): 10.

International Society of Arboriculture (ISA). 1995. Tree-Pruning Guidelines. International Society of Arboriculture, Champaign, IL.

Kalisz, P.J., J.W. Stringer, and R.J. Wells. 1994. Vertical mulching of trees: Effects on roots and water status. *Journal of Arboriculture* 20:141–145.

National Arborist Association (NAA). 1988. Industry statistics. National Arborist Association, Amherst, NH.

Smiley, E.T, D.H. Marx, and B.R. Fraedrich. 1997. Ectomycorrhizal fungus inoculations of established residential trees. *Journal of Arboriculture* 23:113–115.

Sommer, R., H. Guenther, and P.A. Barker. 1990. Surveying householder response to street trees. *Landscape Journal* 9(2):79–85.

Sprugel, D.G., T.M. Hinckley, and W. Schaap. 1991. The theory and practice of branch autonomy. *Annual Review of Ecology and Systematics* 22:309–334.

Summit, J., and R. Sommer. 1999. Further studies of preferred tree shapes. *Environment & Behavior* 31:550–576.

Watson, G.W, P. Kelsey, and K. Woodtli. 1996. Replacing soil in the root zone of mature trees for better growth. *Journal of Arboriculture* 22:167–173.

REPORT ON COMMUNITY

Richard Goodman

Almost precisely a year ago, Rudolph Giuliani, New York City's iron-fisted mayor, collided with a militant gardening populace over the proposed auctioning of 114 community gardens in the Big Apple. In light of that battle, which received nationwide coverage, we thought it time to look at community gardens in general. How long have they been around? How many are there? How are they faring? The skirmish in New York was won, at least for now, by the gardeners, but only with financial help from Bette Midler and the Trust for Public Land. But not every community has a Divine Miss M to rescue its gardens from predatory developers. How can these communities keep the wolf from their gardens?

Community gardens have existed in America for over a century. In 1894, in the wake of a severe economic depression, Detroit's mayor created the first urban plots. The city gave 455 acres of land to 945 families, who grew thousands of bushels of vegetables to sustain themselves. Other major cities began similar relief programs. But the gardens soon went to seed, mainly due to thirsty-eyed real estate developers (has nothing changed?).

The two world wars were halcyon days for community gardens, simply because of necessity. For example, 42 percent of the nation's vegetable production in 1944 was grown in Victory Gardens, which included community gardens, backyards, city parks, and even town commons. There had been efforts in the Great Depression to foster community gardens, but fewer than you might think. Once these disasters became history, so did the gardens—with the exception of Boston's Fenway Gardens, created in 1942 and still going strong.

Community gardens' most enduring leaps forward came in the aftermath of the 1960s. H. Patricia Hynes has ably and gracefully recounted the somewhat esoteric history of community gardens in her book, *A Patch of Eden: America's Inner-City Gardeners* (Chelsea Green, 1996).

In the Northeast, community gardens got a major boost in 1972, when Lyman P. Wood created Gardens for All, precursor of the National Gardening Association. GFA, in Burlington, Vermont, started as a clearinghouse for information on and advice about community gardening all over the United States. A newsletter and books such as *How to Start a Community Garden* and *The Community Garden Book* were published. NGA later changed its focus to children, and its strong youth-oriented gardening program now offers 400 Youth Gardening Grants each year.

From *National Gardening,* June 2000 by Richard Goodman. Copyright © 2000 by national Gardening Association. Reprinted by permission.

The year 1974 was particularly eventful: New York City granted the first lease for a community garden; Ernesta Drinker Ballard, a prominent horticulturist, founded Philadelphia Green, and the Massachusetts Gardening and Farm Act enabled urban gardeners to grow food rent-free on public land.

In 1976, the USDA started its Urban Gardening Program in six cities, to help low-income urbanites grow vegetables. By 1993, 23 cities were involved. Most significantly, in 1979, the American Community Garden Association was born and continues, in its own words, "promoting . . . national and regional community gardening networks—developing resources in support of community gardening—and encouraging research on . . . community greening."

Community Gardening in the Year 2000

What is the health of community gardening today? "The Community Garden movement is growing and growing," declares Sally McCabe, outreach coordinator for the Pennsylvania Horticultural Society and an ACGA board member. "The ACGA has gone from 250 member gardens in 1990 to 900 today." Tom Tyler, ACGA's president, estimates that these 900 gardens represent roughly 500,000 community gardeners. In data from 38 cities in a 1998 survey, the ACGA estimated 6,020 community gardens nationwide, with 2 million community gardeners.

The last five years have seen a 22 percent increase in community gardens: there are now 1,906 in New York; 1,318 in Newark, New Jersey; 1,135 in Philadelphia; 131 in San Francisco; and 44 in Washington, D.C. The greatest threat? Says McCabe, "People are gardening on land that isn't secure. New York City was a wake-up call."

Community gardening is distinctively local. In New York, you find a mayor who, though once an advocate for these gardens, now is indifferent to their destruction. Contrast that with Seattle, where Mayor Paul Schell has mandated that up to five new community gardens be built each year, despite a severe shortage of land. Or Atlanta, where Mayor Bill Campbell is unreservedly supportive of urban gardens.

Positions on land ownership are distinctively local, too. For example, even though Seattle is clearly pro community gardening, the city retains ownership of all the land. Oklahoma City, on the other hand, will deed any vacant lot it owns to a community that wishes to garden on it.

Even the Trust for Public Land's last-minute purchase of 63 of New York's gardens is distinctive (Bette Midler bought the remaining 51). "This is really not the way it should be done in the future," says Susan Clark, public affairs manager for TFPL's Mid-Atlantic Region. The TFPL would prefer that the city's park system had been the buyer. Mayor Giuliani did do the movement a service, though, as Tyler notes: "Controversy is hard, but also good. What the [protesters] did in New York made others aware of community gardens." The fight in New York isn't over, however: in mid-February the mayor had a community garden in the Lower East Side bulldozed.

West Side Community Garden, NYC

One of the most gorgeous, and unusual, community gardens anywhere is the West Side Community Garden, on Manhattan's Upper West Side (spanning 89th and 90th streets between Amsterdam and Columbus Avenues). A quarter of its 16,000 square feet provides vegetable plots for some 83 garden members. The rest of the land is for public use and is devoted to flowers. The garden's most spectacular event is the blooming of some 13,000 tulips in spring, with varieties chosen so that the season lasts six weeks.

The garden survives on its own initiative. To raise money, it holds an annual dinner for which local restaurants supply food. "We usually get 150 people," says Vivian Sanson, who edits the garden's newsletter, "and we charge $50 each." That's more than enough to cover the $4,800 cost of the tulips.

This showcase of volunteerism and hard work is not owned by the city, but by the people who garden it. In 1989, after much effort in front of, and behind, the political scene—something Upper West Siders seem born for—the land was conveyed to the West Side Community Garden, the nonprofit organization that runs the garden. What makes this situation remarkable is that the conveyor of the land could have developed it, as he did neighboring lots, but chose not to. In fact, he was instrumental in deeding it to the WSCG.

Public awareness is extremely important. Although the statistics about community gardening look good, "We've barely scratched the surface," says Sally McCabe. The fact is, thousands of community gardeners do not live in major urban centers and are often ignorant of the help at their disposal. "Gardeners in small communities are isolated," she continues. "They're often starting community gardens in their area for the first time, and they have to convince the town they're not crazy."

A typical case in point is St. John's Organic Community Garden in Phoenixville, Pennsylvania, a former steel town about 25 miles northwest of Philadelphia. Dorene Pasekoff, the coordinator, and her colleagues faced problems in 1990 when they began planning the garden for the town of 16,000. "Basically, in a small town, you're dealing with people who are part-time government officials. We had to spend a lot of time educating them. We had to show them that community gardening is legitimate, that it's part of a nationwide movement, and that it isn't some pea-brained idea we thought up ourselves."

She had a hard time convincing her borough council, and she didn't know about the ACGA. "If I'd had the ACGA information, and the documentation with the official title and the little logos, I could have shown it to them, and it would have all been different."

Pasekoff believes the Web can make the difference. "It's changed everything. I would urge people to go to the ACGA Web site. The start-up information is there. It looks daunting at first, but you can figure it out. Then join their listserv." (A listserv is essentially a discussion group conducted over the Internet.) Here, community garderners can post their questions and hear from those who know the answers.

The future of community gardens may depend on our children's education. McCabe is very keen on California's effort to develop a garden in every school. The results are modest so far, but if it works out, the program could profoundly affect community gardening. As she point out, no school garden can survive without community support—who will tend the garden in the summer and on school breaks?

Vermont is working on a similar initiative. The National Gardening Association is devoting their energy and expertise to educating children about gardening—and backing that effort with grants and a comprehensive Web site.

Even the USDA, which had curtailed its Urban Gardening Program, may be jumping back into the fray. Karen Hobbs, an ACGA board member, who is with the White House Task Force on Livable Communities, reports that the USDA "wants to start a community garden initiative within the department . . . to encourage the development of more community gardens, to start a Web site, to institute an 800 number, etc."

"All politics are local," former House Speaker Tip O'Neill famously said. This—and all its implications—might well serve as the rallying cry for today's community gardeners. Scores of men and women involved in creating and fostering community gardens make it abundantly clear that no really sustainable progress can be made without full local support. There can be setbacks, of course, but amazing things can be done by those with dedication and heart. This much is clear: in the last 25 years, the community gardening movement has developed literature, a store of knowledge, and a network of gardeners the likes of which never existed before. This network is truly a formidable ally for anyone wishing to reap the earth's bounty within city limits.

LIGHT UP YOUR LANDSCAPE

Beth Marie Renaud

Well-placed garden lighting does more than turn night into day. It can transform a dark, mute garden into a lively room to enjoy year-round, whether outdoors in good weather or as a living tableau when you're housebound by rain or snow.

Lighting the landscape can be an expensive proposition if done through 120-volt household current. An economical option is low-voltage lighting. At only 12 volts of power, it is safe, easy to install, and relatively inexpensive to operate: six low-voltage lights use less power than one 75-watt incandescent bulb. The fixtures, wiring, and a transformer to reduce household current are easy to install, and mounted fixtures can easily be moved as plants grow or lighting needs change.

The key to outdoor lighting is deciding what features of your garden to emphasize: a particularly magnificent tree, a colorful bed of annuals, or perhaps a sculpture, a fountain, or an arbor. By selecting which features to light, you create an outdoor room with depth, dimension, and interesting focal points. You can determine what is seen—and not seen. Unsightly aspects of the garden can recede into the shadows while favorite features come to life.

What Do You Want Lighting To Do?

Accent lighting highlights specific aspects of the garden, while task lighting illuminates areas such as a deck, lawn, or entryway where action takes place. Here we'll focus primarily on accent lighting.

To create a balanced and interesting atmosphere in your nighttime garden, use a variety of fixtures and lighting angles, placing fixtures in the foreground, middle distance, and background. However, remember that less is more, and limit the number of light points in each area. Too much lighting fails to focus attention on aspects of the garden you want to emphasize; light can also spill over into your neighbor's property. Hide the fixtures so that only their effects show, not the sources or the glare they can give off. The photos below and on page 44 show a cross-section of fixtures.

Shedding light on a scene can be done in two basic ways: downlighting and uplighting. With either method, the way you direct the light enables you to achieve special effects.

Downlighting is the more common technique. Because the light comes from above, as from the moon or sun, its effects look natural. The most dramatic way of downlighting is to mount canister-shaped lights near the top of a large tree or group of trees. The light that streams down through the canopy mimics diffused light from the moon and casts interesting shadows on the ground. Use three fixtures per large tree, set at different angles, to achieve the most balanced light.

Spotlight a garden bed by mounting a light closer to the ground, on the trunk of a tree, or on the eave of a structure such as a shed or arbor. You can also cast light onto a bed or border by using an upright fixture staked in the ground.

Downlighting is also the most effective way to light a path or walkway. There's nothing worse, when trying to navigate a path, than being blinded by harsh spots of light. To avoid this, use lights with top covers or shades that direct the light out and down but not up. This creates wide circles of softly diffused light that illuminate the ground only.

Uplighting creates a dramatic reversal of natural light, creating striking effects. A canister fixture staked in the ground can send a swath of light upward to highlight a tree's shape, color, and canopy. Also use uplighting for bringing garden sculptures to life, or to show off fountains and arbors. Setting an uplight directly behind an object and aiming the light at a wall creates a distinct silhouette of the object. Bring the light to the front of the plant or object, and you cast its shadow onto the wall instead.

Fixtures

Low-voltage fixtures produce a fairly low level of light, which is desirable for most outdoor lighting needs, but getting enough illumination requires many fixtures: three fixtures per tree for downlighting, one fixture for every 6 to 8 feet of pathway, one fixture per flower bed, and two or three fixtures for a border, depending on its length.

When it comes to selecting low-voltage fixtures, there are numerous manufacturers, so your choices can be overwhelming. Check out Web sites, then visit garden centers or lighting specialists for a taste of your options.

Quality and cost span a wide range. An all-you-need kit with six plastic fixtures and a transformer costs as little as $60 at a hardware store, while individual high-quality fixtures can cost anywhere from $50 to $100 each.

Keep in mind that you get what you pay for. Inexpensive fixtures of lightweight plastic and metal will crack and corrode over time. In contrast, high-quality fixtures made of durable bronze, copper, or cast or enameled aluminum alloy can withstand extreme temperatures and resist corrosion, as can sockets made of porcelain. Most fixtures come with lamps, the industry term for light bulbs, though you may be able to choose a specific lamp for brighter light, a more focused aim, or a broader swath of light. However, be sure to choose only lamps that conform to the manufacturer's specifications in terms of size and wattage.

Installation

Low-voltage lighting—transformer, fixtures, and cable—is relatively safe to install yourself. However, it can be hazardous if the connections at the fixtures are loose or if too many fixtures are connected on one cable. Most equipment comes with detailed instructions for installation, but here are some important things to keep in mind.

Before you install any lighting, check local electrical codes. For instance, some areas allow weather-rated low-voltage flexible cable when installing fixtures in trees, while others insist on metal-clad cable.

Be sure to follow instructions that come with the fixtures, and check with the seller or manufacturer if you have questions. If you are unfamiliar with wiring, you may want to have an electrician do the installation, or check your work.

The transformer converts standard 120-volt household current to 12 volts. Check its wattage capability; most can operate systems using up to several hundred watts. Because there's a limit on the number of fixtures a transformer can power and on the distance the cable can run—about 100 feet—without significant loss of power, divide the lighting load into several circuits of 6 to 10 lights each. The size of the lamps and transformer is also important; consult manufacturer's specifications and stay within required limits. A basic transformer capable of handling six 12-watt lamps can cost as little as $30. One that can support ten 50-watt lamps will cost anywhere from $130 to $350, depending on whether it is electronic or magnetic. To keep the lengths of cable at a minimum, place the transformer where it's central to the fixtures.

Mount the transformer next to an outside outlet, but don't plug it in until you have mounted the fixtures and laid the cable. To add a time clock, plug the transformer into the clock box's receptacle and then plug the clock into the outlet.

Place the fixtures where you want them. Attach one end of the cable to the terminal on the transformer, then connect the fixtures. Some fixtures have a clamp you lay the cable into; other types require that you cut the cable and connect it to the fixture's wiring. Consult manufacturer's instructions.

At this point, turn on the system at night and, if needed, adjust the lights for placement and patterns. The cable can run on the soil surface if plants or mulch will cover it. Or bury it in a narrow trench, taking care not to place it where you might slice it with a shovel. If you have pets that might dig it up, bury the cable about 12 inches deep; otherwise, 6 inches is sufficient.

Once installed, low-voltage systems require relatively little attention, though some routine maintenance will ensure a longer life. To avoid corrosion, remove the lamps once a year and coat the sockets with a silicone-based lubricant. Clean debris out of fixtures, particularly uplighting ones, regularly. Replace dead lamps immediately; otherwise unused voltage will go to the other lamps and cause them to burn out faster. Lastly, prune vegetation around fixtures to maximize light output. The exception to this may be with tree-mounted fixtures; small branches growing in front of these lights help to diffuse their bright light and cast intriguing shadows on the ground.

FOOT DR.:

HEALTHY FEET MAKE HEALTHY GREENS

Dr. John Perry

Overuse, repetitive motion and improper shoe gear are the primary causes of lower-extremity injuries.

Injuries fall into two categories. The first, acute trauma or accidents, makes up 20 percent. The second, repetitive trauma or "overuse," is much more common at 80 percent. Prevention of these injuries can result in huge savings in terms of time and rehabilitation expenses.

One pair of boots or sneakers, replaced regularly, can help avoid arch pain. Proper forefoot fit can prevent mortons neuroma, a pinched nerve in the toe area.

Regular behavior changes, like changing socks and shoes at lunch time, can reduce the chances of getting athletes feet. This tip can be especially helpful with the double whammy of morning dew and automatic sprinklers.

Superintendents frequently ask "Hey, doc, what shoe should I wear?" The answer is "It depends."

A 200-pound person with high arches needs more cushioning and less support. A 150-pounder with flat feet needs more support with a rigid heel counter and less flexibility in the sole.

Regardless of the individual's dimensions, all golf course footwear should have the following:

- ❑ rigid heel counter;
- ❑ roomy toebox;
- ❑ firm outer sole, at least an inch thick that flexes only at the ball;
- ❑ breathable upper material.

I don't recommend particular brands because shoe styles change too frequently.

Shoes should be changed every day to dry out and replaced every three months. It may seem expensive, but will result in cost savings by reducing lost time and injuries.

"But I'm doing those things, doc, and still have pain," patients sometimes persist. Don't despair.

If it's ankle, knee, hip or back pain, there could be a simple solution. It begins with a thorough physical exam and medical history. Then, using computer gait analysis, a diagnosis is made.

On the golf course, if the soil is too acidic, then growing problems occur. In lower-extremity biomechanics, when the leg and foot roll in excessively (over pronates) the ligaments are stressed excessively, causing shin splints, knee bursitis and/or low back pain.

The solution? Foot supports, commonly known as orthotics or arch cushions. The over-the-counter variety are an inexpensive first step, but often aren't overly effective.

If fatigue and pain persist, custom-made orthotics may be the answer. They can be soft, semi-rigid or rigid, depending on the individual's needs. They fit directly into the shoe, where they better control abnormal pronation motion and any resulting pain.

Just as good soil conditions cause grass to flourish, good mechanics result in reduced stress and fatigue. That makes for fewer injuries and a happier grounds crew.

A Meaningful Mix

Terry Ostmeyer

For a great many superintendents, the routine of golf course management has increasingly included family time.

As the job of course maintenance evolves into a more agronomically precise task, it's also becoming more visible, more competitive and more pressure-packed. Superintendents, in turn, are becoming more desk-bound with the diverse management chores of planning and delegating.

Daily opportunities for head superintendents to examine the golf course up close and personal seem to be fewer and farther between. For most, the best time to check on the state of their playing fields comes somewhere beyond the boundaries of the so-called work day—early morning, evenings and weekends.

So Little Time

That Superintendents put in long days and nights is a given. It comes with the territory and always has. But this time-worn and time-honored job description does beg a question seldom addressed on a large scale until recently: What's a family man or woman to do?

Indeed, spending quality time with one's family, especially the children, would appear to be an ominous void in the life of a superintendent. For many, maybe so. But for others, the void is filled as best it can be by sharing some of their work life with their families.

Of course, few, if any superintendents have family members in tow during the prime working hours of the day, but managers traditionally spend countless "off hours" and weekends checking their golf courses and catching up on office work. It's not uncommon for those with children to bring them along. It's their version of a day at the park with the kids. More important, it's making up for a lot of lost time at home.

Although old-timers in the business say having the kids along at the course now and then is nothing new, it certainly appears to be a practice that has grown tremendously. Besides the obvious increases in golf courses and superintendents, more and more families are marked by both parents working, requiring many more hours of day care and shuffling of schedules by parents. Another factor is that there are more single parents today, and the ramifications of that and of holding down a virtually 24-hour, seven-day-a-week job are often even more intense.

Family Ties

In many cases, superintendents' kids tagging along with their dads and/or moms to the golf course has some traditional roots of its own. Many of today's golf course managers are second- and third-generation superintendents who were brought up on the course.

"It was a big highlight for me when dad brought me to the course, and I think I can see that in my kids now," one superintendent told *GCM*. "During the heart of the season, course maintenance is dawn-to-dusk coverage all week, so time with the kids is precious."

Wives of superintendents have strong feelings about the issue as well. Mrs. Mark Blest (Honey Brook (Pa.) Golf Club), for instance, noted that the favorite activity of their three small children is to go to "Daddy's golf course," which she says helps integrate their world with his.

Amy Cox, wife of Ted Cox, superintendent at Running Fox Golf Course in Chillicothe, Ohio, points to a more intrinsic need for their two children to be with their father at work now and then: "My husband works ungodly hours, especially in the summer, and if it wasn't for the time we do spend with him on the course, we would never see him."

Pamela's Dilemma

The subject of superintendents' children at the workplace surfaced late last fall on the GCSAA Web site's "Talking it Over," a members-only message board. The posts were in response to a missive on the site from Pamela Smith, superintendent at Blackberry Patch Golf Club in Coldwater, Mich., a nine-year GCSAA member and the mother of Connor Smith, who will celebrate his first birthday in June.

Smith is a single mom, and balancing the dual responsibility of full-time superintendent and parent of an infant son has been a challenging feat over the past several months—to say the least—and a problematic one as well.

"Connor spent up to 50 hours a week in day care during our busy times last summer," Smith notes. "I've got to fit in there somewhere, and I've brought him to the course in the evenings to check on things like pumps, on weekends for office work, and even on my actual days off when something has come up that requires me to be there."

This exhaustive slice of life suffered a setback in mid-October when Smith was out one weekend morning with Connor touring the Blackberry Patch layout in a golf car. The twosome was noticed by the facility's general manager. Already beset by one of the course owner's worries about liability and another course employee who had brought a child to work, the GM wrote a letter to Smith stating that employees' children were not allowed during "scheduled" working hours.

Smith's reaction was varied, but initially much of it centered on the perception of some of her superiors that superintendents are governed by set working hours. Still, more than that, she says now, the situation was kindled by a lack of communication among managers, limited knowledge of and experience in golf course management outside the superintendent and maintenance staff, and ownership's concerns over liability and employer/employee relations.

The significant response to Smith's dilemma and her subsequent posts on the Web seeking input from her colleagues ran heavily on the supportive side and nearly all of that from superintendents who also bring their children to work at times.

Degree of Difficulty

Although Smith maintains her situation was primarily the result of communication and policy-making problems, there were some superintendents who wondered aloud on the Web site whether the real issue might be gender.

One veteran superintendent, Dale Walters, CGCS at Royal Palm Country Club in Naples, Fla., doubts those implications per se, but adds that, as a woman, Smith is a departure from the norm in the male-dominated ranks of golf course managers.

"I would think that for a woman to mix child care and our profession as opposed to most of the men who do it would be not only quite different, but probably quite difficult," says Walters, a 24-year GCSAA member.

Citing a decade of experience and many dues paid along the way, Smith dismisses the gender issue in favor of what she calls greater concerns—lessons to be learned.

"I think the substance of this issue isn't a child at work. We all have proved this is a norm," she wrote following a flood of posts supporting her stand. "The issue is informed policy making from management and a better knowledge of our profession. I've read a lot of posts here regarding job displacement. I can't imagine another career where you put as much into facility planning for the future and have it taken away not because of your product, but rather a 'personality' conflict with management."

At Blackberry Patch, the issue was on simmer in mid-February. Smith and her GM had resolved their differences, and the next step was for superintendent, GM and ownership to sit down and develop a policy that addresses the concerns of all.

"I can understand their worries over liability, and I can see their point concerning all course employees, because, where do you draw the line?" says Smith. "But being part of a family is so important, and what's the joy of this job we have if you can't share it with your family?"

The Paper Trail

To be sure, bringing children to the golf course workplace is both a legal and a management issue, says Jules De Coster, lawyer, superintendent and golf course owner from Monticello, Mo.

"The child is an invitee, and given the same reasonable care as others on the golf course, ownership is not liable if something happens," says De Coster, who owns and operates Fabius River Links in Montecello and also is the prosecuting attorney of Lewis County, Mo. "It's common sense on the superintendent's part—you don't have a kid on a mower or a tractor or playing in the maintenance shop . . . common sense."

De Coster, who also runs a risk-management consulting service for the golf industry, RIMAGO Co. Inc., is a strong advocate of superintendents working out a policy with ownership on the issue of children at work. Such a document, he notes, should include these basic points:

❏ The child should be a certain age, preferably in line with an all-course policy.
❏ The child must comply with all course restrictions the same as all other employees.
❏ The policy should apply only to family members of the superintendent.
❏ The child should have health and accident insurance coverage.
❏ The policy must be enforced.

De Coster adds that the question of who the overall policy covers is very important because other employees in a golf course operation have become increasingly aware of the issue.

Communication and Responsibility

The responsibility of the superintendent for the actions and safety of his or her children in the workplace was strongly voiced among the Web posts by a veteran Oregon course manager and 15-year GCSAA member who asked that his name not be used for this article.

The member noted that he asked his employer if his children could accompany him on the job during off hours. After checking with the resort's insurance company, which said the facility was covered, permission was granted.

Pat Kriksceonaitis, superintendent at Essex Country Club in Manchester, Mass., says he signed a waiver of liability with the private club so his daughter Samantha, now 5 1/2, and son William, 16 months, could be with him on the course during evenings and weekends. He notes that his wife, Lowell, is a horse trainer with hours that rival those of a golf course manager.

"It's a simple document. The club's not liable, and we all understand when the kids can be there and when they can't," says Kriksceonaitis, a six-year GCSAA member. "It's simple, straightforward and everybody's happy with it. I know that without it I don't think I could function as a superintendent or a parent.

"Both kids love going to the course. The oldest realizes working is part of life. For me, they help make the long hours go faster, and I don't seem to worry as much."

Personal Feelings

Still, not all superintendents think having children in the workplace is a good idea any time of day or night. Keith Pegg, CGCS, who was superintendent at Fircrest Golf Club near Tacoma, Wash., for almost 20 years before getting into golf course development in Asia, said in his Web post that staff's children have no business on a golf course unless playing the game within the course's rules.

A GCSAA member for 26 years, Pegg told *GCM* he was surprised at the number of today's superintendents who do bring their kids to the course after viewing the number of Web posts responding to Pamela Smith.

"Still, I feel quite strongly about their not being there," Pegg says. "God forbid if they should get hurt. That should be everyone's concern."

Owning Up

Although liability is a common thread found in the issue, it nevertheless appears that many golf course owners have bigger legal worries.

"Our world is very litigious. . . . It seems like every time you turn around a golf course owner is being sued for something," says Anne Lyndrup of the National Golf Course Owners Association.

Lyndrup says at first glance NGCOA's main concern with children in the superintendent's workplace would be the liability question. Yet, she adds, the association—which has more than 2,800 members representing 3,900 courses, of which 70 percent are public-access venues—actually has not addressed the issue much to this point.

"It seems to be more of an issue for those parents who are trying to work and raise a family—like the superintendents," Lyndrup says.

As a matter of fact, Lyndrup, who is the director of NGCOA's Get Linked to Play Golf Program, is heading up the association's "Take Your Daughter to the Course Week" April 24–28, which includes "Take Your Daughter to Work Day" on April 27.

A Little Showing Off

Like fellow veteran Pegg, Walters was somewhat surprised at the number of superintendents who share their workplace with their children these days, and even more so by the passion for doing so displayed by many of them.

But for Walters it was a pleasant surprise, for he is one of them. All four of his children have enjoyed the experience. Two are grown, but Walter's 13-year-old son Joshua, and 12-year-old daughter Charity, continue to join him and revel mostly in the wildlife found on the Royal Palm CC property.

"They love it, and I think it's great," he says. "We're environmental caretakers, and that's a big deal nowadays. The kids can see for themselves that it's really so."

SPECIALIZING IN FARMERS' MARKET SALES

Betsy Hitt

While Betsy and Alex Hitt aren't growing the crops they thought they'd be in 1981, they're still making a living from the soil.

Rather than growing berry crops, Peregrine Farm is setting a prime example of how small growers can succeed with specialty cut flowers. The company produces fresh cuts on 2 acres and sells primarily at a local farmers' market, but that's enough to make the Hitts happy. For succeeding with specialty cut production and farmers' market sales, Peregrine Farm has been named a **GM**PRO Innovator.

Alex, with a degree in soil science from Utah State University, and Betsy, with a degree in forest recreation from the same school, settled in Graham, N.C., with plans to sell pick-your-own blackberries and raspberries.

"After three years we determined that our business plan just wasn't working," Betsy said. "We started clearing more land for vegetable production and a few years later began to grow a few flowers. They immediately gained my attention."

Betsy began concentrating more on fresh cut crops. It wasn't long before the couple became familiar with a then-new organization, the Association of Specialty Cut Flower Growers. By attending this group's annual conference, they met small specialty cut growers like themselves and learned more about the trade.

Four Main Crops

Larkspurs, lisianthus, zinnias and sunflowers make up the bulk of Peregrine Farm's cut flower production. About 40 other cut annual and perennial crops are also produced, but in much smaller quantities.

Most of the production is outdoors, but the Hitts do have four movable polyethylene Atlas Greenhouse Systems houses (total 3,000 square feet) that can provide protection from inclement weather. They produce mostly fresh cuts, although there are a few dried crops such as chili peppers.

"We just can't get the prices for dried material to justify the extra labor and handling they require," Betsy said. "We couldn't sell anything dried until fall anyway. It would take that long to process. And a lot of things can happen until then, like mice or rehydration."

Focused on the Farmers' Market

About 80 percent of the Hitts' sales take place 15 miles away at a farmers' market for Burlington/Chapel Hill/Durham area growers. The other 20 percent are mixed bouquets sold to area grocery stores.

The farmers' market is open 35 days a year on Wednesday evenings and on Saturdays, when crowds are bigger. The Hitts operate a 25-by-18-foot booth that costs about $265 a year in rental fees. The couple's turn per square foot annually is about $100.

Of the 80 vendors at the market, almost all sell cut flowers at one time or another, Betsy said.

"There are about four companies like ours that are serious about flowers. The rest may just be picking lilacs out of their yard when they're in bloom," Betsy said. "But flowers have a very high visibility at our market and that's a good thing."

With so many vendors offering flowers, the Hitts make sure their cuts are the highest quality possible.

"I'm probably too picky about what we sell," Betsy said. "Pricewise, I don't pay attention to what other people are doing. On occasion someone will come up to me and say that someone else has what I do for 5 cents a stem cheaper. I'll tell them as politely as possible that they should go back and buy them.

"I know if they're coming back here after seeing flowers cheaper somewhere else then my quality must be better."

Rotation and Cover Crops

The Hitts rarely have to deal with disease or insect problems and credit this to crop diversity, crop rotation and cover crops. By planning production practices far in advance, no insects or diseases are allowed to get out of control.

When cover crops are tilled under, a portion is left standing to provide a shelter bed for beneficial insects. This year, the Hitts plan to grow some plants—including carrots, dill and cilantro—around their operation to attract more beneficial insects.

Every year, each growing section is soil tested for fertility, which can result in as many as 24 tests annually. Cover crops are planned based on what will be planted in the beds next. If an early spring crop is scheduled, then crimson clover or oats are planted as a cover crop. If a later summer crop is planted, then rye or hairy vetch is used.

"You try to get the maximum biomass out of a cover crop possible," Alex said. "Rye and hairy vetch provide a lot of mass and nitrogen, but you can't kill it in the spring. You have to wait until it has matured or you'll never be able to get rid of it."

Cover crops are sometimes sandwiched between two cut crops in one season. For instance, larkspur can be followed by soybeans or cow peas in mid-June and replanted again with a late cut crop in September or October.

Increased No-till

The Hitts are using no-till agriculture more, especially with vegetable crops. With no-till, a cover crop is planted, allowed to mature and then mowed down. This kills the cover crop and the vegetable or cut flower crop can be planted directly into the cover crop remnants. If a cover crop is mowed too early, a herbicide must be applied to kill it.

Straw mulch is commonly used to reduce weeds around cut crops. Landscape fabric is used also, especially for perennial cut crops. The fabric is used with the crops during the first year and removed before plants re-emerge during the second season.

SHATTERING THE GLASSHOUSE CEILING

Jennifer D. White

The road to equality, no matter the industry, doesn't come without bumps, detours, stereotypes and stories of discrimination

"There will be differences in any profession that was once male dominated," says Sandy Cruise, a woman who has earned the coveted position of head grower at one of the country's premier greenhouses. In horticulture, quality and hard work precede reputations, and in the last 30 years women have shown that they can grow plants and run businesses right alongside men.

We wanted to profile a few inspirational women in our industry to show how they got started and where they are today. But we found more success stories than we have pages! So we want to acknowledge that these four individuals represent a multitude of talented, hard-working women in horticulture.

Louise Tramontano, co-owner, The Garden Spot and The Greenery, Center Moriches, New York

There's no simple way to approach an article on Louise Tramontano. She's the former president of Bedding Plants International (BPI), a grower, business owner, mother and computer whiz. She has two daughters who are rapidly making their own mark in horticulture. In fact, when talking about their retail center, The Garden Spot, Louise points out that "there are no men here—it's all women." Her daughters, Kristin Comeau, 29, and Kimberly Tramontano, 27, have soaked in their mother's ambition. "Both of them now are taking over a huge chunk of retail." She adds, "My greatest pride is that two women are rising up and excelling."

But Louise also points out that it's a family affair—her husband, Tony, runs the wholesale side of the business, The Greenery, growing nearly all the plants for The Garden Spot. Louise and Tony jumped into horticulture in 1973 when they purchased The Greenery.

Neither had a horticulture background, and they relied on the good-hearted nature of greenhouse owners. "We received a tremendous amount of help," says Louise,

---

citing friendly neighbors and fellow growers Russell Weiss of Kurt Weiss Greenhouses and Jack Van de Wetering of Ivy Acres Inc.

Louise admits that in the beginning it was difficult to garner recognition and respect because she didn't have a horticulture background. And even though she was responsible for the initial phases of production, customers were sometimes wary to approach her, rather than her husband, for crop information. But times have changed, and over the years Louise has proven her leadership capabilities, serving on the Agricultural Committee, the Farm Bureau's Women's Committee and the local school board. She says her 12 years on the school board—6 as president—prepared her for her role as president of BPI.

With Jack's backing, Louise got onto the BPI board in 1990 and in 1996 she was the first female grower elected president. "There was no glass ceiling there," she says.

Louise knew BPI was headed for a rocky period before she stepped into the position, and, during her tenure, in 1997 the association was forced to cancel its annual conference. "That was probably one of the two most difficult things I had to do in my life," she says. But Louise was backed by a supportive BPI board and a family that kept the plants growing while she tended to her BPI responsibilities. Yet despite the difficulties, Louise says, "Being the first woman president of BPI will certainly always be a highlight of my career."

Since passing the torch on to current president Henry Huntington, Louise has been able to turn her attention back to The Garden Spot. Beside her are two equally driven daughters, who proved they were capable of taking over when Louise's time was diverted by BPI.

Kristin is the one responsible for creating The Garden Spot in 1988 when she grew and sold a crop of mums and used the profits to buy a greenhouse. It wasn't until 1993 that Louise and Tony decided to get involved with retail rather than compete in the wholesale bedding plant market. Kristin earned her associates degree in horticulture and a bachelors in marketing, with retail as a minor. Kimberly has also made her way back home and put her artistic skills from the School of Visual Arts in Manhattan to work, remodeling the store for spring 2000 and taking responsibility for buying everything except plant material. She travels the country collecting new trends and ideas. Today, 90% of their business, more than $1 million, comes from retail.

Life for Louise is about change. "I hope I'll never settle down, never stop growing," she says. She recently received the equivalent of a bachelor's in computer science and is now responsible for all advertising and Web site business. The Garden Spot, a Home and Garden Show Place, took one of the top three awards for Web pages at the Las Vegas TrueServe Show.

"I think you can do anything you want," Louise says of the expanding roles of women in the industry. "My advice is to always be open to change, always move ahead and don't be afraid of it."

Virginia Walter, interim department head, Environmental Horticultural Science, Cal Poly, San Luis Obispo, California

Virginia Walter grew up in a family that was forever gardening. "We didn't do anything else," she recalls with a laugh. "I gardened with my mom and dad. When my dad was pouring concrete, I was there. I never learned that one was men's work, the other women's."

You might say that Virginia has paved the way for women, being the first female to hold many titles and honors—including president of Pi Alpha Xi, the national honorary society for horticultural students. "But I've never looked at it that way," she says, explaining that because she was at the front of the wave, it was only natural for her to fill many of these positions first.

For the last 26 years, she has been guiding top horticulture students through the environmental horticulture program at Cal Poly. She travels all over the world to speak about U.S. floriculture. She was awarded a prestigious Fulbright scholarship and spent six months teaching in Zimbabwe.

When Virginia was earning her horticulture degree at The Ohio State University, she was one of the only women in the program. But today she doesn't see gender being a big issue in our industry. For several years, Cal Poly has had at least 50% women in their programs.

Over the years, Virginia has sent out wave after wave of horticultural students into the workforce. And they aren't average students either. In 1999, she coached her flower judging team to their fifth national championship in 10 years in the Intercollegiate Flower and Plant Evaluation and Floral Design Competition. When asked why they've been so successful, she explains that her students have the opportunity and the facilities to be well prepared for competition. But Angela Lopopolo, a student under Virginia for the last 2 1/2 years, adds they also have a great coach. "She's very stern makes you know your stuff," says Angela, "but she's very supportive. She's a very strong individual; she knows horticulture in and out."

"I'm always adjusting my teaching techniques to keep the attention of different generations," says Virginia. And while she possesses a depth of knowledge, Angela points out that she doesn't just lecture: "She gets you to think on your own."

"My students are like my family," says Virginia. She opens her home to her students and dedicates herself to coaching, serving as an advisor for Pi Alpha Xi and other agricultural projects. She also serves on several committees and is in the process of developing a CD-ROM for her greenhouse environment class—all while guiding her students into top-notch careers. "I'm very proud of where they're going and what they're doing."

Anna Ball, president and owner, Ball Horticultural Company, West Chicago, Illinois

"I wasn't really raised to go into the business," says Anna Ball, president of Ball Horticultural Company—a business that has been in the Ball family since it was founded in 1905. "My brothers were raised to go into the business, but I wasn't, which was normal back then."

Anna was nearly 30 before she started her career in the seed room and began to work her way up the buying area of Ball Seed. It was an experience, she says, that was very beneficial and gave her a lot of respect for the employees.

Even though Anna worked hard and attended night school for five years to earn her MBA from Northwestern University, she admits that her family name gave her a big advantage. "I'm here because of who I am," she says candidly, "not because I worked my way up, even though that's what I did."

What's it like being a mother and running a company that includes Ball Seed Co., PanAmerican Seed Co., Ball FloraPlant, ColorLink, Ball Superior Ltd. and Vegmo Plant B.V.? Time is always an issue. As an owner, she has control over what she does, but the responsibilities of running Ball also dictate much of that time. "I just believe in doing everything passionately and with energy. If you're working, you work with energy and you throw yourself into it. And when you're with your children, you just throw yourself into it," says Anna. "I'm what they classify as a single mother. It's tough, but it's satisfying, and it's fun."

Anna isn't the first to admit that there can be disadvantages to being a woman. "Some women have to take care of their parents, their kids, and work at the same time," she notes. "But, I think it's great." She loves being a mother, being the primary caregiver and running a company.

She says she doesn't think about the disadvantages or even the fact that she's a woman spearheading a leading company in international horticulture. "I don't believe anyone can believe I don't think about it, but I don't! I just think that I have a business to run and I've got to do the best that I can and try to put everything into it."

And when asked how she balances all this, she replies, "Balance is really boring. If you try to balance everything, then you don't do anything either really well or with energy. So I sort of swing!"

When Anna was first beginning to do work in Asia in the early '80s, there were times when she wasn't treated entirely seriously, when people would only look at the man sitting beside her while they talked. "But that's really long gone," she says. "Partly because everybody I meet knows me and knows the company."

While Anna can say that her family name got her where she is today, her leadership and dedication keep Ball growing successfully. "I've always played the role of—I guess I got it from my mother—keeping the wheels going," she says. Whether she's talking to customers or an industry gathering, it's clear she has the enthusiasm to lead into the future.

For young people entering the industry, she has two pieces of advice: "One, work your tail off. Two, follow your heart, not the money."

Sandy Cruise, head grower, Metrolina Greenhouses, Huntersville, North Carolina

Growing. It's long hours, hard work and a lot of responsibility—especially if you're working at Metrolina Greenhouses in Huntersville, North Carolina. When Sandy Cruise started as a section grower in 1986, Metrolina was just 17 acres. For the last four years, they have been adding an average of 5 to 6 acres per year, bringing them up to 70 acres today.

Metrolina supplies much of the mass market, including Kmart's new Martha Stewart line, and with so much expansion, it's been necessary for Sandy to adjust with time. In the beginning of her career, being a grower was about focusing on crop culture and scouting for pests and diseases. Today, however, Sandy finds being a grower has more to do with production numbers, scheduling and being ready for sales.

"It's taken a lot of adjusting. The preplanning is much more important than it was 10 or 15 years ago," she says. Trying to fit everything the salespeople want to sell into their existing greenhouse space is often one of the bigger challenges.

Being flexible has become a necessity—learning to hold plugs a bit longer and then getting products shipped out as quickly as possible when spring rolls in. And as a mother of two children, ages 4 and 7 years old, she must bring that flexibility into her home life as well. Both Sandy and her husband work as growers at Metrolina, and when spring season hits, it's always a squeeze to spend as much time as they'd like with their children.

"That has probably been the most difficult in the last few years—juggling between home and work," Sandy admits. "But I don't feel like my kids have suffered by my working," she states, adding that it helps that she and her husband work together.

Sandy confesses that she, too, has run across her own gender-bias challenges, especially at her first horticulture jobs following graduation from Kansas State. She wasn't always taken seriously at first.

"I've always felt I've had to prove myself a bit more to gain people's respect," she says. But, "Anybody who chooses horticulture knows the long hours and hard work that comes with it." When you're doing your job and putting the effort in, your work is noticed and you gain that respect. Her advice to young women is to just "be willing to work as hard as anyone else."

Only a year after Sandy started working as a section grower at Metrolina, the head grower left and she swiftly worked up into his position. Since then, she's proven to be adept at growing with the times. "My expertise is that attention to detail," she says. "When I stand back and look at it, I'm responsible for millions of dollars of plants. That's huge." And as Metrolina continues to speed on into the future with high-tech automation, computerized systems and quality plants, Sandy is proof that, male or female, a good grower is a good grower.

Jumping on the Floral Cart

Monica Humbard

Floral industry experts aren't the only ones recognizing the potential of uncharted territory in the floral industry. Big names like Martha Stewart, Hallmark, and Sears, Roebuck and Co. are carving their own niches. Note their high-profile efforts because they not only increase competition in the industry but also project an image directly to floral consumers. On the positive side, if the efforts of these companies are successful, they could help expand floral's piece of the pie in the gift market.

Sears, Roebuck and Co.

Although Sears brings to mind home appliances for most people, in recent years it has redefined its image. In February 1998, the company spun off a home decorating and remodeling store in Denver, called The Great Indoors. Included was a test floral department. Last November, Sears opened a second location in Scottsdale, Ariz., covering 133,000 square feet and housing an expanded floral area.

In The Great Indoors floral department, customers can buy containers and permanent flowers by the stem, similar to a Pottery Barn. They can either design their own arrangements to coordinate with home-decor items purchased in The Great Indoors' other departments or they can seek the advice of a staff designer. Staff designers will construct the arrangements as well. Around the holidays, the store offers classes such as "Decorating Your Table" and "Floral for the Season."

The test store in Denver started with limited space for the floral department and concentrated on high-end silk and dried stems and finished products. Customer feedback revealed a need to broaden the selection to include lower price-point options as well. While the price options have expanded, however, Michelle Pratt, decorative accessory buyer for The Great Indoors, says quality has remained a primary concern. She points out that the company is not trying to compete with the hobby and craft stores.

Fresh flowers were tested in the Denver store as well. "It just didn't work for us," Pratt says. "The customer didn't think of us in that way, so we pulled it out."

Pratt considers the floral departments successful. "It's become a real impulse category. It's driving a lot of sales," she says.

New Great Indoors stores will open this summer in Dallas and in the Detroit area at the end of 2000. The company also is planning to open a second store in Denver. Pratt doesn't expect the floral program to change much in the near future, but she says the store will continue to monitor customers' opinions and needs.

Hallmark

Unlike Sears, Hallmark is not a first-timer in the floral industry. This is the company's second shot. Its previous attempt was not successful. O. Stanley Pohmer Jr., Pohmer Consulting Group, Minnetonka, Minn., recalls that the first time Hallmark actually included floral inventory in its stores, involving displays and dedicated coolers. He says operational issues (replenishment/ inventory management, store execution, etc.) led Hallmark to abandon its initial floral efforts.

Rachel Bolton, a Hallmark spokesperson, says the focus and distribution process are different this time. Consumers can order flowers through the Hallmark Web site, *www.Hallmark.com;* by calling an 800 number; or at a kiosk in Hallmark Gold Crown stores in five test cities. The kiosks offer pictures of the available products and a phone line directly connecting consumers to the distribution center in Memphis, which ships the orders to customers via FedEx. The kiosk program was launched last October in Albany, N.Y.; Charlotte, N.C.; Portland, Ore.; Sacramento, Calif.; and Columbus, Ohio.

Flower orders are shipped from U.S. and international growers to the distribution center. "That cuts down greatly on the time it takes to get flowers to the consumer," Bolton says. She says consumers should receive them within two to six days from when they are cut.

Hallmark flower orders are guaranteed and include an assurance that the flowers ordered will be the same flowers delivered. "Hallmark has ensured the flowers are properly cared for along the way as far as temperature and climate control," Bolton says. "If the recipient or sender is not delighted, Hallmark will either send a replacement bouquet or a refund."

Bolton says Hallmark research revealed that freshness is one of the most important issues to consumers. In a Hallmark test study, nearly 90 percent of Hallmark flower recipients were very satisfied with both the appearance and condition of the flowers at arrival and seven days after. Bolton says research revealed a 30-point drop in the satisfaction level after one week with traditional floral delivery systems.

Along with distribution changes, Hallmark has revamped both its product and its packaging since earlier attempts.

"Hallmark research has indicated that European countries are more particular and demanding about their flowers—both with quality and the type of flowers," Bolton says. "They are very much a part of their lifestyle. Hallmark is providing for consumers here the look and feel of the kinds of bouquets that are popular in Europe. These don't have that 'arranged' look."

Flower choices will change with the seasons, but 15 different bouquets are available, ranging in price from $39.95 to $69.95 plus shipping.

The European-style bouquets arrive in maroon gift boxes with crystal vases and wire-edged ribbon. Flowers are packaged in a moisture-retention pack, placed in a vase and wrapped in tissue with a Hallmark gold seal and a full-sized greeting card. A cold pack also is placed with the arrangement in a double-insulated box. Because they are prearranged, the consumer can remove the flowers from their moisture pack, fill the vase with water and drop the flowers into the vase. A preservative packet and care-and-handling information are included.

Bolton says Hallmark has done a lot of research on brand extension and has learned that this is an area where consumers will "give permission" for Hallmark to affect their lives. "We know that people rely on Hallmark to help express their feelings and emotions and establish and nurture relationships. Flowers are an important way to do that."

Hallmark is testing different advertising venues in the targeted test cities to determine what is most effective. "What we learn from the experiences in these cities will be used to make sure the best possible methods are used," Bolton says. She adds that the program is expected to launch nationwide in Gold Crown stores after Valentine's Day.

"With Hallmark's gift emphasis," Pohmer says, "they could be a factor that affects traditional florists, 800 providers and other e-commerce floral retailers. I don't see them taking many sales from the mass market."

Martha Stewart

Martha Stewart followers probably could have guessed it was only a matter of time before she expanded into floral sales. Last October she began offering consumers grower bunches, including roses; greenery; vases and accessories via her Web site, *www.MarthaStewart.com,* and an 800 number. Consumers can purchase 25 roses, available in 12 colors, for $52 and 50 for $98.

Since the program was launched, a Flower of the Month Club was added. Consumers can order a three-month ($138), six-month ($238) or year ($438) subscription for themselves or someone else. Each month they receive a different type of flower. For example, subscribers will receive 30 mixed-variety daffodils in February and five mixed-variety hydrangeas in September. Orders are being filled by USA Floral Products, Miami, Fla. Consumers choose to receive their flowers, which are delivered via FedEx Priority Overnight, either the second or fourth week of the month. The service is limited to the United States.

Besides standard shipping charges, customers pay an additional $5 for every shipment per single address. Care-and-handling information is included with all floral orders and also is available on the Web site.

In addition to her Web site, Marthas Flowers also will be featured in her gift catalog and her consumer magazines.

"At present her assortment offering is somewhat limited, but I think it will expand over the short term once they get over some of their conservative approaches to the program," Pohmer says. He also noted that the product is pricey but believes consumers consider her name "worth some incremental price value."

"I believe that this program will reach more new floral consumers than cannibalizing existing trade venues," he says. "She has a franchise with consumers, and many current nonfloral consumers will try her program simply based on her name strength. Her program may eat into some of the other Web-based floral retailers and some traditional florist sales but will have little impact on the supermarket/mass-market sales venues."

Despite repeated attempts to contact the company, no one from Martha Stewart would comment on the company's entry into the floral market.

"It's interesting to note that both (Hallmark and Martha Stewart) are disintermediated models," Pohmer says. "They're trying to shorten the distribution chain, bypassing wholesalers and retailers. This is a good news/bad news scenario. The good news is that they are taking cost and time out of the process that can benefit the consumer and/or their bottom lines; the bad news is that they take on 100 percent of the shrink and operational issues themselves. If they have the expertise in-house to manage this process, it will improve their chances of success. If they don't, it could be a disaster for them."

SERVICE ON THE SPOT

Dan Anderson

House calls providing on-farm oil changes and maintenance work save time, money

Grimmway Farms, Bakersfield, Calif., produces 18,000 acres of carrots and vegetables each year using a half dozen D-7 and D-8 Caterpillar crawlers, 13 Caterpillar Challengers and dozens of wheeled tractors.

Their Cats often run 20 hours a day, six days a week, putting as many as 2,500 hours a year on individual units. Through it all, employees never change a drop of engine oil or replace a single fuel or air filter.

All maintenance, from oil and filter changes through belt and valve adjustments, is handled on site by their Caterpillar dealer under a customer service agreement (CSA).

The CSA, similar to those consumers buy for appliances, is a new offering from Caterpillar dealers. Individual dealers for other equipment companies offer self-generated agreements, but Caterpillar is the only company with a formalized program. Such service agreements are commonplace among Caterpillar's industrial customers.

Under a CSA, the dealer contracts to provide regular maintenance on site, at prescribed intervals, for a negotiated fee. Parts, fluids, filters, labor and travel costs are incorporated in a single fee spread over the life of the contract.

"Having a mechanic do the maintenance and look at the machine on a regular basis has paid," says Carl Voss, a Grimmway Farms manager. "If he catches and repairs a small problem before it becomes a major breakdown, he has saved us not only the cost of major repairs, but the $35 an hour we'd have to pay to rent a crawler while our machine is in the shop."

Full-service CSAs are not cheap. Proponents say the initial sticker shock of CSAs points out the "hidden" cost of good maintenance.

"If you spread the cost of a CSA over 2,000 engine hours, the cost averages $1.50 to $3 an hour," says Paul Rice, product support manager with Quinn Company in Fresno, Calif.

"If a farmer figures what oil, filters and other maintenance items cost him on a per hour basis and charges himself for the time it takes to do professional-quality maintenance, he'll find that doing it himself isn't as cheap as it seems," he says.

Customer service agreements also address the environmental hassles of disposing of waste oil, used antifreeze and dead batteries.

Holt Equipment in San Antonio, Texas, spent $150,000 outfitting five Peterbilt-based service trucks so they can safely and legally transport bulk and waste oils. Equipped with cell phones and pagers, the trucks are on call 24-hours a day, seven days a week.

"We do service work at night, in the morning, on Sundays—whatever it takes to keep the farmer running," says Don Easterwood, agribusiness manager for Holt.

Forty-eight of the 50 Challenger tractors Holt Equipment sold in recent years had a CSA incorporated into part of the purchase agreement.

"Farmer's attitudes are changing," says Easterwood. "This generation of farmers is more like businessmen. They're busy dealing with labor, crop prices, banking, soil fertility, weed control—they aren't interested in being mechanics."

Emmet County Implement, a Deere dealership in Estherville, Iowa, takes a different approach to on-farm maintenance. Their On-Farm Preventive Maintenance Program provides free labor for changing engine oil and replacing filters in the field. The customer pays only for the oil and filter, plus $1.50 to dispose of waste oil.

Each visit includes a 66-point inspection of the tractor or combine. Up to 200 customers a year participate in the program. If the inspection turns up problems, the customer must approve further repairs at normal service call rates and parts-counter prices.

"We ask that they call a half day or more in advance so we can schedule the oil change," says service manager Mike Dalen. "We do a lot of oil changes while farmers are breaking for lunch or supper." The program is especially popular during harvest.

"There's a lot more that can go wrong on a combine," says Dalen. "Changing the oil gives our guys a reason to check each machine at least once during harvest. We've caught quite a few problems before they became big and expensive."

ORGANIC PREMIUMS

Darrell Smith

Premium crops pay more but challenge management

Ten years in South Africa doing maintenance at a mission hospital gave Ray Berry a fresh perspective on farming. There, he says, "organic farming was just regular farming."

Turning to organic production after returning to Norris, S.D., helped Berry, his wife, Gail, and sons, Loren and Brent, survive the 1980s. "The premiums kept us in business," he says. As a bonus, soil quality has improved and erosion has been eliminated.

Now organic crop rotations are insulating Berry's 3,000-acre operation from the low prices that are buffeting conventional farmers. His 1998 barley crop netted $117.60/acre; sunflowers netted $131, and millet about $100, while conventional winter wheat was losing money. He also raises spring wheat, corn, alfalfa, flax, buckwheat, soybeans, oats and grass seed—all grown without herbicides or fertilizer.

Berry pastures 120 brood cows. He rotates all his fields through several years of pasture, harvesting grass seed and hay before the cows go on. The cows and grass improve soil quality and help control weeds.

Last year, Manska intermediate wheat grass was Berry's best crop. Including the value of seed, hay and grazing, it netted $413/acre.

Rotation is critical. Organic is not an easy way to farm. "We spend all winter deciding what crops to grow," says Berry. Not all are in demand every season. When the Iron Curtain fell, East German buckwheat crowded the U.S. crop out of Europe. Berry has three years' production in storage, which he will have to grind for feed. Meanwhile, the European millet market has turned sluggish because of the rising value of the dollar.

Since weed control and fertility depend on rotations, Berry must fit the crops he wants to grow into the proper sequence. Barley, a cool-season grass, for example, works well after millet, a warm-season grass. Sunflowers, a broadleaf, do well following grasses—either small grains or sod. "Because they are planted later, we can till before planting to control weeds," says Berry. Soybeans, a broad-leaf and a nitrogen fixer, are ideal ahead of small grains.

Planting time must be considered because, in organics, it becomes a weed-control tool. "Quackgrass and pigeon grass germinate from mid-May to mid-June, so we want to plant sunflowers before or after that period," says Berry.

Even seed size must be considered. For example, flax is a bad fit after millet. If the millet volunteers, the flax must be cleaned so hard that some of it will be lost.

Marketing decisions also are more complicated than those for conventional crops: Not every crop is contracted, and terms of contracts vary. Some prices are quoted at the farm and some at the terminal. Dockage is higher than for conventional commodities and, for some crops, the grower bears the cleaning cost.

Even when a crop is contracted, the producer must store it until the buyer wants it. Berry has 48,000 bu. of storage. "There are many hoops to jump through," he says, "but that's true of any identity-preserved crop."

Tricky growth. Organics may offer opportunity for a small number of producers willing to manage intensively and do their homework. If you think you are one of them, you can learn about opportunities as Berry did, by attending organic agriculture meetings. Then you need to line up a certifying agency; certification takes three years.

Berry expects demand for organic crops to increase, but cautions that the markets for many crops are almost saturated now. "We could stand 100 new growers per year, but 1,000 per year would be a problem," he says.

"The one crop that fits anywhere is grass," he adds. "Grass seed is terribly touchy to harvest. You have to dry it, and it's difficult to handle. But it kept us in business several years. Besides seed, it provides hay and pasture. Any seedsman can tell you about varieties that will work in your area."

Growing Organic

John Dillon

With a soil map in hand, Will and Judy Stevens went looking for a place to live. They found their home ground in a fertile layer of earth in Shoreham called Nellis loam, deposited by the last glacier.

Their Golden Russet Farm is in southern Addison County, which has traditionally been cow country. Modern metal-roofed barns dot the landscape, and the flat fields of feed corn seem like a slice of Iowa transplanted to Vermont. But at Golden Russet the only hint of anything bovine is the manure piled in an open bunker, making compost for the season ahead.

The Stevenses' 82-acre operation is one of 187 certified organic farms and processors in Vermont—a more than tenfold increase since 1985, when the state had just 17 organic operations. In the last five years alone, the number of organic farms in the state has almost doubled, according to the Northeast Organic Farming Association of Vermont (NOFA), the Richmond-based organization that certifies farms and processors as organic.

The Stevenses moved to Shoreham in 1984 from Monkton, a town south of Burlington that even then was feeling the pinch of Chittenden County suburbia. The couple was looking for cheap land and a place where farms still thrived. When the real estate agent pitched a likely prospect over the phone, they'd whip out the soil map and see if it was worth driving down for a look.

"Soil was everything for us. Soil and, as it turned out, community. The more we looked, the more we realized we wanted to be in an agriculture community," Will says.

The couple were pioneers in Vermont's newest wave of organic farming, a technique that harnesses the biological systems of the land and shuns chemical fertilizers or pesticides. The organic farmer has to pay close attention to pest control and soil health. Soils are replenished with nutrients derived from cover crops, compost, manure or other natural sources. Crop rotation is crucial, both to ensure healthy humus and to prevent bugs and parasites from gaining a foothold.

Will, a broad-shouldered former blacksmith who studied studio art at the University of Vermont, speaks of organic farming as more than a system of soil management. "After you do it for a while, it becomes a way of looking at life—and a way of living," he says. "After you've done it for three or four years, you look at the whole system organically, how doing one thing a certain way affects everything way down the line, and it forces you to plan ahead. It involves the end-purchaser and the community you're part of. It's not just about production."

At first, the organic movement wasn't market-driven; it was born of back-to-the-land beliefs and a growing environmental awareness. In the early 1980s, when Will would make the run from Monkton to the Burlington farmers' market early on Saturday morning, he didn't even label his vegetables as organically grown.

"This tells you how far we've come," he says, while taking a break from tending the broccoli on a sunny July morning. "At the time, we felt the perception of organic produce was wormy cabbage. We didn't want people to see the word *organic* and be put off by that and then to go on down the line. We wanted people to come to our table, to see good-looking produce and buy it . . . and if they learned it was organic in the process, that was a bonus."

These days, consumer demand has far surpassed any worries about blotchy brassicas. And as the Stevenses' experience shows, the grower now reaps the bonus because he or she can charge more for organically raised food.

Over the last decade, organic farming has moved from the fringe into the mainstream to become the fastest-growing sector of U.S. agriculture. Americans spent an estimated $3.5 billion on organic foods in 1996, according to industry figures. Up-scale natural food chains such as Whole Foods market and Bread and Circus thrive in metropolitan areas by tapping into the growing demand for organic food.

Will and Judy Stevens are now veterans in a business that has remade the face of Vermont farming. As the dairy industry struggles with chronically low prices, organic agriculture has emerged as an important force in the state's farm economy.

NOFA requires farm operators to detail current production plans and demonstrate that the soil has been kept free of chemical pesticides and fertilizers for three years. Notarized affidavits and annual compliance checks are used to ensure organic practices are followed. Annual application fees range from $150 to $550, depending on gross sales.

The Vermont farms certified through NOFA represent about 15,000 acres, with annual sales of around $25 million. Many raise vegetables, but organic dairy farms have also seen phenomenal growth, going from just three in 1993 to 38 last year. Vermont's own Organic Cow, a Tunbridge company founded in 1990, was sold in 1999 to Horizon Organic Dairy Inc., the nation's largest distributor of organic milk. Although organic methods can be expensive, for dairy farmers the rewards are particularly sweet. Organic Cow pays roughly $23 per hundred pounds of milk, almost twice as much conventional milk handlers.

Consumer demand is driving the tremendous growth in farm numbers, says NOFA director Enid Wonnacott. Nationally, organic sales have grown by about 20 percent a year since 1989, fueled by the public's desire to eat healthy foods, avoid pesticide residues and support environmentally friendly agriculture.

Wonnacott believes the market is far from saturated. "We intend to do a lot more consumer outreach," she says. "We're doing a lot more work with schools, to get schools to purchase locally. We think there are huge places for expansion."

Vern Grubinger, director of the sustainable agriculture program at the University of Vermont, likens modern agriculture to two trains running down diverging tracks: One track runs toward larger, energy-intensive farming operations that rely increasingly on hormones, chemicals and genetically modified crops; the other leads to organic farms connected economically to local communities.

The public clearly wants farms to provide wholesome food and to tread lightly on the environment, he says. As evidence, Grubinger notes that in 1997 the U.S. Department of Agriculture received a record number of comments—a blizzard of 280,000 letters, e-mails and postcards—from people opposed to a government proposal to weaken organic certification standards. The USDA has since issued relatively strong rules.

Organic farmers are also reinventing the market so they can capture more of the money spent on food. "Many farmers are bypassing the distributors and processors that gobble up their profits," says Grubinger, "and instead are marketing directly to consumers, or making their own value-added products."

Like the Stevenses, many who came into the business early chose organic production for lifestyle reasons or to protect the land. Paul Harlow of Westminster, named by NOFA and the state Agriculture Department as the 1998 Sustainable Farmer of the Year, says he left chemically based agriculture in 1976 for personal reasons.

"It was health [concerns] and what I thought was good for the land," he said. "I could see that in 25 or 100 years from now my farm and the land were going to be much worse for wear."

Harlow markets much of his produce through the Deep Root cooperative, a Waterbury-based association of about 15 growers who pool their efforts to land large accounts with wholesale buyers in Boston, New York, New Jersey and the Washington-Baltimore area. Harlow's farm, with annual sales of about $500,000, is the largest co-op member.

As his production increased, Harlow invested in machinery to ease the back-breaking labor of hoeing weeds and pulling late-season carrots out of near-frozen ground. He toured European organic vegetable farms in 1989 and adopted some of their techniques, including the use of a flame weeder that scorches pest plants but leaves the crops intact. A later visit to California organic farms inspired the purchase of more specialized cultivating equipment used to control weeds.

Harlow has converted his family's two-century-old dairy barn into a storage and packing center. A huge insulated room keeps the peas, beans and other produce chilled to 33 degrees. Nearby, a large "hydro cooler" washes and chills heads of lettuce by bathing the plants in frigid water. The storage rooms also allow him to sell produce like winter squash right through until spring, providing needed cash flow through the lean months.

"We spent a lot of money. It made things tight financially. But now . . . we can take the time to fine-tune it and concentrate on making a profit," he says.

For Richard Wiswall, owner of the Cate Farm in East Montpelier, the breakthrough to getting his operation securely in the black came about when he went deeply into debt. In 1985, Wiswall borrowed almost $200,000 from a state-backed loan program to buy out the original partners in the farm. The loan process pushed him to pay close attention to his balance sheet and to study exactly which crops made money and which were losers.

"We were almost forced to sharpen our pencils and look at how to make serious money to pay for the mortgage," he says.

Wiswall keeps accurate time sheets for every stage of production—the planting, weeding, harvesting and packing each crop requires. He knows that he or his employees should be able to wash and pack three to four cases of spinach an hour, or assemble 150 to 200 bunches of cilantro. With the financial records and time sheets as his guide, he has decided to grow more medicinal herbs—selling some through a Web site—to capitalize on the expanding demand for natural medicines.

Wiswall, Stevens and Harlow are veterans of the early days when a small fraternity of organic growers shared expertise at NOFA workshops and spent hundreds of hours drafting certification rules to regulate the industry. Now a younger generation has opened new markets, often through "community-supported agriculture," a system in which a customer pays up front for a season's worth of vegetables.

Burlington farmer David Zuckerman, who in the winter months cultivates votes as a state legislator, founded a farm last year in Burlington's intervale, a broad swath of fertile Winooski River floodplain tucked inside the city limits.

Zuckerman worked several years on a farm in Grand Isle, then at the Stevenses' Golden Russet Farm before striking out on his own. The eight acres he tends in the Intervale supply food for 72 people who subscribed in advance. Zuckerman says his customers are enthusiastic about CSA, farmers' shorthand for community-supported agriculture. His small farm is one of three CSAs in the Intervale.

"There's competition, but the demand potential here is huge. On my opening day, I had 40 people sign up," he says.

The CSA system has also helped Will and Judy Stevens, but in their decade and a half in Shoreham, they have weathered lean years, drought and competition from California growers. Nothing prepared them, however, for the mid-summer morning in 1997 when a crop-dusting airplane banked low over their neighbor's cornfield. Some of the herbicide drifted from its intended target and settled over their rows of growing vegetables. The fields were devastated. Onions wilted, the growing buds of broccoli turned yellow and died. Their prized heirloom beans were history.

Golden Russet Farm was laid to waste by the kind of agribusiness practice Will and Judy had worked years to avoid. But it could have been worse. The crop duster's insurance covered the loss, and extensive soil tests later that summer showed that the herbicide had broken down and could not be detected in their fields. With that evidence in hand, NOFA allowed the couple to resume production the next year.

Will says he didn't even think of suing his neighbor, who ordered the spraying of the crops. They're good neighbors and his family and the Stevenses' kids go to school together. "I just don't have the time for that," he says. "I got my payment [from the insurance]. I was satisfied."

(He quietly donated $2,500 from the settlement to a NOFA fund that provides disaster relief to farmers. The money came in handy during the dry summer of 1998.)

On a fall weekend following the spraying disaster, hundreds of friends and strangers showed up for a benefit to honor the farmers and their commitment to the land. It was an affirmation of the community Will and Judy Stevens were looking for and found back in 1984.

"I don't know how I could begin to repay that kind of show of support," Will says. "There were old friends and people we had never met who just knew of our situation and wanted to come down and chip in. . . . Our feeling was that we were just a couple of woodchucks who wanted to do what we wanted to do and then we had this interruption. But then all of a sudden it became apparent how important this role we were playing was."

A Day for Trees

Arbor Day, which falls on the first Friday in May in my state, isn't much observed anymore. Perhaps it has lost its appeal in Vermont because three-quarters of the state is now forested. The opposite was true in 1885; citizens eagerly supported the state's first Arbor Day, when three-quarters of the state had been clear cut.

Then-Governor Samuel E. Pingree decreed that "the love of Vermonters for trees and groves should show itself," and it did. The village of Proctor alone planted 2,000 trees. Yet Vermont's numbers paled beside Nebraska's, where 1 million trees had been set out on the very first Arbor Day celebration in 1872. By 1887, 600 million trees had been planted nationwide; today, every state and even several foreign countries recognize the holiday.

Arbor Day was the invention of J. Sterling Morton, an ambitious new graduate of the University of Michigan. Only 22 when he migrated to the Nebraska Territory in 1854, Morton had headed west looking for "agricultural possibilities," despite warnings that the land "wouldn't raise white beans." In fact, the land was rich and fertile, and Morton planted trees as well as corn on his farm, which he had claimed by squatter's right. While serving as president of the Nebraska Board of Agriculture, he proposed an official day for planting trees. Calling the settlement of America "a diary of destruction," Morton lobbied for an arboreal bureau to "act as a signal station does upon a storm coast, and warn the race . . . from danger to its very existence which shall come from nonattention to forestry—too much activity in cutting down and too little in planting out of trees."

Arbor Day quickly evolved into a patriotic holiday as well as an occasion to set out huge numbers of trees. Towns were encouraged to hold "proper exercises, the recitation of brief passages from English literature relating to trees, songs about trees sung by the children, addresses and planting of trees, to be named for distinguished persons. . . ." By 1890, only communities south of the Mason-Dixon line lacked a Lincoln oak.

An Arbor Day manual published that year consists of 456 pages of suggested poems, songs, and readings. Despite a warning—"Caution: Do Not Make The Program Too Long"—one program includes nine musical selections, two prayers, nine recitations, three essays, one address, one declamation, and one dialogue, plus a reading of the Arbor Day law, a vote for a state tree, the appointment of a tree committee, and the naming and planting of trees.

From *Horticulture, Gardening at its Best,* May 2000. Copyright © 2000 by *Horticulture.* Reprinted by permission.

Programs like that may explain the demise of formal Arbor Day festivities. But planting trees still deserves recognition—an annual day to offset the paving of America. At the risk of being called old-fashioned, I planted a pin oak in my yard in 1988, in honor of Governor Madeline Kunin. Then I took a tree-planting break while a Republican ran the state. Now I'm looking for a spot for a Howard Dean mountain ash.

If you want to add a note of authenticity to your commemorative tree planting, look into American Forests' nonprofit Famous and Historic Trees program (8701 Old King's Road, Jacksonville, FL 32219.

ANIMAL INDUSTRIES

OVERCOMING OBSTACLES TO
INCREASE BEEF DEMAND

Adam Jones

Well, we've got trouble right here in the beef industry And that's trouble, oh yes! We've got lots and lots 'a trouble. Trouble, with a capital T, and that rhymes with B, and that stands for beef! Yes Sir, we've got trouble with Beef Demand.

The United States Beef Industry employs over one million people and produces nearly 25% of the world's beef supply. Cattle production, the largest segment of the agriculture economy contributes $153 billion dollars to our nation's economy. Almost half (45%) of the U.S. cattle businesses have been in the same family for 50 years.

There is nothing wrong with the United States beef production, but beef consumption is a totally different story. Consumption of red meat has decreased 15 pounds per person over the last 20 years. Although this year we have seen a small turnaround in consumer demand, what can we do to keep this trend heading in a favorable direction?

To market any product, the consumer needs to have faith that the product they are eating is safe, affordable and healthy. The cattle industry has worked hard to insure that our product meets all three of these. Still it's difficult to change people's preconceived ideas, so how can we overcome these obstacles to increase beef demand?

The obvious answers are public education, convenience, and so-called "Branded Beef." The beef industry is currently working on these ideas. In addition, value-added foods help the consumer prepare healthy, tasty meals in minutes. Branded Beef products are now a reality, helping to insure consistency of quality. Certified Angus Beef, U.S. Premium Beef and other similar programs are starting to become recognizable to the American consumer as products of superior quality.

Meat cookery education is being addressed right at the meat counter with the Meat Case Simplification program. Stickers are put on the packages of meat with cooking instructions. These solutions to increasing beef demand are in effect and the industry is using them to help solve the on going problem. Nevertheless, rather than looking to the past solutions maybe we should stop what we're doing and look to the future consumers . . . Children!

According to information supplied to the Youth Strategic Planning Task Force by the Geppetto Group in New York, "The generation of kids now five to 18, the sons and daughters of Baby Boomers, rivals the Boomers size and buying clout."

A survey conducted by Nickelodeon says that 88% of children today prepare their own breakfast. This same survey also stated that one-third of all kids prepare their family dinners. Children three to four years of age recognize brand names and between the ages of five and seven, they are going to supermarkets and purchasing items by themselves.

Children are unique consumers:

They are purchasers—By the end of this year kids will spend more than $300 billion.

They are influencers—in the food category alone kids influence the spending of $88 billion. Ever gone grocery shopping with a child?

They are the future—James McNeal, professor of Marketing at Texas A & M University says, "As a future market, kids have the most market potential, for they will eventually buy all goods and services. If a business nurtures kids as future consumers, they are more apt to like that company and prefer its products and services once they get old enough to buy them on their own."

So what will the marketing motto for the next decade be? "If you don't have a kid's product GET ONE! And maybe it's time the Beef Industry got one!

Kids are very brand oriented when purchasing goods. We see evidence of this by what kids buy. Nike tennis shoes, (are they really better that those Payless specials?) Gap Jeans (honestly they don't fit any better; they just have a classy label.)

To kids, beef is NOT a "brand." As producers, we are aware of the marketing benefits of branded beef, but we are not thinking about kids when developing a beef brand. Kids want to know that a product is made for them. They want something that they can connect to. Something they can identify.

Of course, we know that creating a Branded Beef product is a slow and expensive process. But maybe we could create a personality or character that signifies "kid approved" products. Just look at what Tony the Tiger did for frosted corn flakes. This beef character would be wherever beef products are kid-appropriate, whether it is the meat case, or the ready to serve and frozen items.

When we (the beef industry) develops new ads, we need to make them "Kid-Fun." Our access to reaching children is quite vast through cable television, magazines and radio ranking high on influencing teen's marketing decisions. Whether you admit it or not, teens are sophisticated consumers; they want products of superior quality. Nonetheless their overriding motivation is to buy fun products. We as the beef industry need to convey the message in our ads that the reason they are having fun is because they are eating beef. (A hamburger picnic at the lake with all your friends; a party at your house where everyone is making and eating beef tacos.)

Youth are surprisingly health conscious. From 1998 to 1999, 10% more young men said, "The protein found in red meat is important to building a healthy body." Children have a perception that they should be health conscience, but that nutrition is boring. To help them see that beef is good for them, they need to see their heroes eating beef. Preferably the heroes who are their age or a little older, like teenagers.

Maybe even the local high school track star rather than the national basketball player. Someone they really could be like if they would just eat nutritious, delicious beef!

As I have presented, the cattle industry has some troublesome obstacles to overcome to increase beef demand. There are many ideas in the works including branded beef, scientific studies and education about the healthy attributes of beef. One of the ways I see to lessening these troublesome obstacles to increased beef demand is to extensively market to children. Develop a branded beef that appeals to kids. We should use advertising to show people of their own age reaping the benefits of eating beef, and remember, kids just want to have fun. So why not have fun preparing and eating Beef.

Well, we've got a future. A future like this industry never had. We've got Influencin', Consumin' Youth, and that starts with a capital Y. And that rhymes with buy; and that means beef. Buy beef, Yes sir.

TOMORROW'S BEEF PRODUCER

Kindra Gordon

Expect more beef alliances in the next decade, even though almost 60% of beef producers who responded to an exclusive Intertec Publishing survey say they don't plan to build formal partnerships.

That's an indication, says North Dakota State University extension livestock economist Harlan Hughes, that the beef industry could segment into two groups: quality-oriented producers and commodity producers.

Despite all the talk about vertical integration, 62% of survey respondents say they aren't currently involved in formal partnerships with suppliers or customers; 59% don't plan such partnerships.

"I think the future is bright for the group willing to build partnerships with alliances," Hughes says. Those who continue as commodity producers "don't want to change and don't realize the world's changing around them; they'll need to be low cost to survive."

Even if the industry segments, most producers are optimistic about beef a decade down the road. Regardless of gross sales or age, over 40% of producers across all income levels say that they have an excellent to good outlook for the industry's future; 36% are neutral; and 20% are leery of what's ahead.

Hughes says alliances and new product development are key factors in producer optimism for better times ahead.

A third of those surveyed believe development of new beef products will have the greatest impact on the future success of the industry in the next decade.

What concerns them. Low prices, environmental regulations and competitive markets top the list of producer concerns. Given the economic picture of the last five years, that's not surprising, says Hughes.

Finding viable markets also concerns producers; 61% worry that the export market won't be able to handle excess beef if the domestic market keeps shrinking.

The U.S. Meat Export Federation's Philip Seng says otherwise. He points out that 96% of the world's population is outside the U.S. and says "the market is there."

Federation projections through 2005 put exports at 1.8 million metric tons (they reached 1 million for the first time in 1998). "We've been laying the groundwork in

---

DO YOU EXPECT TO BUILD MORE FORMAL PARTNERSHIPS IN THE FUTURES?	
No	58.5%
Yes	41.5%
Total	100%

opening markets," Seng says. "Buying power is there, trends are favorable, and consumption is increasing internationally."

To corner that market, Seng believes more commitment to the international marketplace in marketing and policy is needed. "Producers need to be active in their respective associations and aggressively took at export markets," he says.

What kind of cattle will be raised? Seventy percent of survey respondents expect the number of breeds to stay the same over the next 10 years; just over 20% expect fewer breeds.

Dave Nichols, Bridgewater, IA, concurs with the majority, but cautions that there may be fewer seedstock providers. "By 2010, I look for 100 seedstock providers, in 2020 only four," says the Angus, Simmental and Salers breeder. Of those, he expects more will be offering three- and four-breed composite lines.

"There's a tremendous merger mania going on," Nichols says. "It's naive to think that what's happening in other industries won't happen to beef."

He's upbeat about the future. However, one reason beef hasn't gone to corporate breeding establishments, Nichols says, is because breed associations have done a good job of tracking EPDs and getting information back to producers.

Embracing technology. The Internet could play an important role in the beef industry, respondents say.

Jim Gibb of eMerge Interactive, a Web-based company that provides information management, technology and e-commerce to the animal industry, says the Internet

What is your attitude for the future of the beef cattle industry?	
Excellent	2.3%
Good	42.1%
Neutral	36.2%
Poor	19.5%
Total	100%

The Next Decade of Dairy

Like all of agriculture, the dairy industry will see fewer and larger operations over the next decade, says University of Wisconsin dairy specialist Bob Cropp.

Cropp says those changes are happening rapidly. "Over the next five years nationwide the number of dairy operations could be reduced by one-third."

The dairy industry is already seeing a trend toward 1,000–2,000 head operations. That has been facilitated by production and information technology that allows the management of a large number of cattle, Cropp says.

He adds, "This doesn't mean we won't still have some viable smaller producers. But to survive, they will need to be low cost."

Cropp forecasts that the dairy industry will continue to be strong in the West, particularly Idaho and California. He believes the industry could see growth in the upper Midwest; and the Southeast will continue to struggle due to high production costs.

With increasing herd size, Cropp says larger operators will be buying more services such as heat detection and breeding assistance, as well as consultant advice.

Information technology could have the biggest impact on the industry. Cropp says that by using computer technology, producers will be better able to monitor feed costs and herd health, which in turn will increase productivity.

In the future, Cropp believes membrane technology, a process that takes the water out of milk, may help expand and extend the marketing options for milk.

will become a common channel for distributing beef information and electronic commerce in the industry.

Less than one-third of the beef producers responding to the survey say they now use the Internet, but half say they expect to do so in the next decade, about an 18% increase.

Gibb expects tremendous growth in technology that can capture information in real-time, analyze the data and return it quickly to producers.

"Agriculture is in an era of precision management. It's going to be extremely critical for producers to have daily information" to improve production efficiency, carcass information and consumer-defined palatability.

"It allows producers to manage with binoculars instead of through the rear-view mirror," Gibb says.

Some producers are already eager for such information, and "those that are more likely to be profitable in 2010 will utilize real-time information to make decisions." He also sees Internet auctions as a "huge medium for marketing feeder cattle and seedstock."

Other technologies they hope to be using that aren't common now include sexed semen (18%), electronic ID (26%), and embryo transfer (20%). Only 3% think cloning will be part of their operation in 2010.

The beef business. No matter what direction the industry heads in the next 10 years, most respondents (71%) anticipate staying in the business.

Inadequate income and retirement were the two top reasons they would call it quits. The average age of respondents was 54. But beef producers under 35 years old were unanimous—100%—that they'd be in the beef business 10 years from now.

A 40-year-old Alabama producer predicts: "Things will change more in the next 10 years than in the past 20-25 years. Those unable to change will have a hard time. The best educated and most informed will be on the cutting edge."

Sizing up Rivals to the North

Tyler Kelley

The United States has every advantage to become the No. 1 pork exporter in the world. You produce a low-cost, high-quality product in a country with free-trade agreements and no unreasonable restrictions.

Don't look now, but you're not the only one who can say that.

In the pork industry, looking to the north is like looking in a mirror, as Canada vies to compete with the United States for the title of the world's No. 1 pork exporter.

"The bottom line is the United States and Canada are increasingly becoming one market," says Chris Hurt, Purdue University agricultural economist. "The Prairie Provinces are becoming really competitive and they're going to be a long-term force."

Canadians raise about 20 million hogs, well short of the U.S. totals. However, the population is smaller than in the United States, so Canada has been exporting a larger percentage of its production longer than the United States has. The leading market for Canadian hogs and pork has always been the United States.

While a lot of Canadian pork has been flowing south across our borders, the big impact has come from Canadian live hogs. In 1999, 4.1 percent of hogs slaughtered in the United States were from Canada. Ms number was about the same as in 1998, when Canadian imports contributed to the U.S. slaughter capacity problems.

Many Canadian hogs come in as feeder pigs and are fed out in the United States. In 1998, 2.7 million market hogs came from Canada, with an additional 1.5 million pigs weighing less than 110 pounds. Market hog imports fell to 2.1 million in 1999, but imports of lightweight hogs rose to 2.1 million, points out Hurt.

On the reverse, Canada's border has opened to any state with a pseudorabies status of Stage IV or V. That should have allowed some U.S. hogs to move north, but Canada's international trade considerations have slowed this process.

"Australia won't allow imports from Canadian packing plants if those plants slaughter U.S. hogs," says Nick Giordano, trade counsel for the National Pork Producers Council. "This is a bogus sanitary standard for pseudorabies, and it's a clear trade barrier."

The United States opened the border further by recently dropping the countervailing duty on Canadian hogs. The duty was originally put in place in the 1980s to account for subsidies that Canada was paying its producers. Once those payments ended in 1997, the protective duty came off. While NPPC favored keeping the countervailing duties a possibility, the U.S. government dropped the duty.

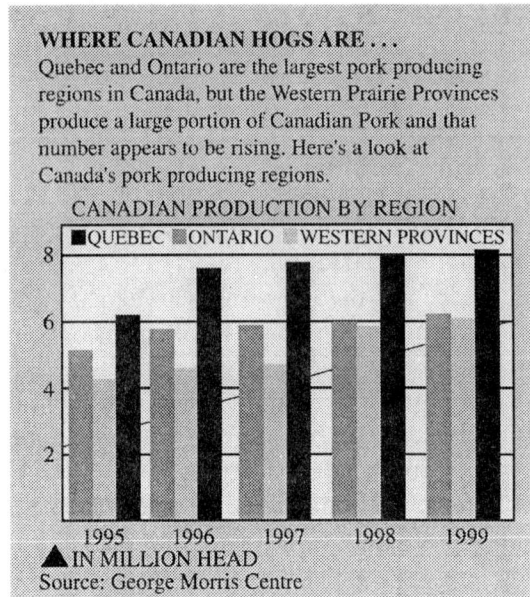

WHERE CANADIAN HOGS ARE . . .
Quebec and Ontario are the largest pork producing regions in Canada, but the Western Prairie Provinces produce a large portion of Canadian Pork and that number appears to be rising. Here's a look at Canada's pork producing regions.

CANADIAN PRODUCTION BY REGION

■QUEBEC ■ONTARIO ■WESTERN PROVINCES

▲IN MILLION HEAD
Source: George Morris Centre

Shipping U.S. hogs to Canada could become a more viable option thanks to an increase in packing capacity. Maple Leaf Foods opened last year in Brandon, Manitoba. When the plant becomes fully functional and runs a double shift it will be able to kill more than 2 million hogs annually.

"Manitoba will need more than 8 million hogs for its new and proposed plants to make it, and last year Manitoba only killed 5 million hogs," says Giordano.

Larry Martin, CEO of the George Morris Centre, an in dependent agribusiness think tank in Guelph, Ontario, says it's reasonable that Canada's annual capacity will reach 20 million hogs, but that Canada has never slaughtered that many. Long-term, it could offer another market outlet for U.S. hogs, provided trade barriers are resolved. Certainly fewer Canadian market hogs will head for the United States.

"People are sending fewer weaned pigs south and are starting to put up more finishing barns," says Martin. "Others are expanding." Also, some British and Dutch pork producers who have been displaced by restrictive production regulations and ailing profits are immigrating to set up shop in Canada.

The two countries have more fairly balanced trade agreements as of late. Since negotiating the U.S /Canada Free-Trade Agreement in 1988, Canadian pork exports to the United States declined, while U.S. exports to Canada rose to more than 45,000 metric tons, according to Giordano.

While the United States may be Canada's No. 1 pork export market, it doesn't mean the Canadians aren't a major player in other markets as well.

"Canada is neck and neck with the United States as the No. 2 pork exporter behind Denmark," says Giordano. "Canadian exports have been increasing like ours."

The United States currently dominates the Japanese fresh/chilled pork market, but the Canadians sell frozen and chilled pork to Japan. During fiscal year 1999, Japan bought 72,163 tons of U.S. pork, while Canada exported 42,543. Denmark still leads with 96,393 tons.

Clearly, Canada is eyeing Japan, which is part of the force behind its slaughter plant building boom.

More questions surround China's market potential. Based on population and pork usage, it offers the richest export prize ever. However, culture, transportation and refrigeration factors will limit the upside somewhat.

The United States has a pending agreement with China that says China will accept U.S. pork from any USDA inspected plant. The Canadians don't have that same deal. Of course, the Canadians aren't dealing with a Congress that's dragging their feet on the U.S./China trade status.

The United States was able to send its first official pork shipment to China this spring.

While the Canadians pose a serious threat to the United States' long-term pork export goals, this cloud may have a silver lining.

"The Canadians have a quality product and are viable competitors," says Giordano. "We meet with them frequently and there's an opportunity for the two countries to work together to promote free trade."

The same similarities that make Canadian pork production a rival to U.S. production also make the Canadians trade allies to the United States. For example, Canada introduced the zero-for-zero initiative that the United States embraced at the last World Trade Organization negotiations. They are the only two countries to have signed the initiative at this time, but will continue to work to persuade other nations.

Ultimately, the take-home message is that you can't afford to slip in terms of quality, safety or cost efficiency of the pork you produce. If you drop the ball, someone else—most likely the Canadians—will be there to pick it up and run with it. At the same time, if Canada and the United States can work together to open up even more markets for pork exports, the future should be profitable for both countries.

Perception vs. Reality

Jane Messenger

===

An on-farm odor/environmental assessment can help bridge this gap for you and your neighbors.

Public perception and the reality of pork production don't always match, especially when it comes to environmental issues. This isn't an easy battle, but there are a lot of options available to put the odds back in your favor.

Just ask Marshall, Mo., producer David Bentley. No one has ever complained about odors from his operation, but the potential is there. Bentley has 15 neighbors within a mile of his facilities, with at least half of these, including a public golf course, closer than a half mile.

In July 1998, Bentley decided to take advantage of the National Pork Producers Council's On-Farm Odor/Environmental Assistance program. The program is free, all it takes is some of your time and willingness to make your operation more environmentally friendly.

Here's how it works. Contact your state pork producer association and request Form A. Use this document to provide information about your operation, including contact names, location, a general description of your production system, neighboring land use, biosecurity requirements, hog mortality disposal, manure management and land application practices.

Return the completed Form A to your state association. It will then head to NPPC where an assessment team will be assigned. Within two weeks, assessors will schedule a site visit at your farm. One assessor will represent the private sector, the other is from a non-regulatory government agency or university extension service.

You will accompany the two members during the assessment. They will complete a detailed site checklist and give you a verbal report of their observations when the visit is completed. Within two weeks, the assessors will send NPPC a detailed written report, which they also send to you, outlining the key risk areas for odor and water quality impacts. The report includes contact names and other information that you can use to obtain any design, management or cost-share assistance that may be necessary to correct the risk areas.

If your operation has a severe or high-risk challenge, one of the assessors will do a follow-up phone call or visit within 90 days.

In high-risk cases, a third-party verifier also will conduct a random follow-up of the team and their assessment to verify that the program is identifying environmental risk areas and what procedures, if any, you are taking to correct problem areas.

All information resulting from the audit is kept confidential.

"The improvements that we needed were more like insurance, to keep us from having a spill," explains Bentley "If we had a spill, it would cost us plenty with the Department of Natural Resources and in terms of the press and our relationship with our neighbors. By doing an assessment, we can keep something bad from happening. It's worth it."

The outcome of Bentley's assessment was positive, and he shared some of the recommendations that the assessors had for his operation.

Recommendations

1. Put a manure management plan in writing. Bentley has always kept track of the gallons of manure he pumps, where he spreads manure and soil test results. But, with only he and his father running the operation, he didn't have a formal plan in place. He now has a written plan.
2. Install a pump-down stake in the lagoon to measure when it is full. Although Bentley's experience told him when the lagoon was beginning to fill and automatically pumped it down two feet each spring, he agreed to install the measuring stake.
3. Install a "Do Not Enter" sign in the area where he pumps the lagoon. He hasn't put up a sign, but hopes that people use common sense anytime they're around the pumping area.
4. Replace one of the wooden covers where the pit flush lines intersect. Bentley acknowledges that the cover was rotting. He has replaced it.
5. Fence off the lagoon. This was a recommendation that Bentley did not implement because it is too costly and not a priority at this time.
6. Keep predators out of the compost pile. He deals mostly with coyotes and wild dogs. Bentley tried putting up an electric fence, but it didn't work. The best he can do is to check the compost pile daily. If some of the areas are dug up, he adds more sawdust.

"Producers are correcting about 50 percent of the identified problems," says Dan Uthe, director of the National Pork Producers Council's National Swine Information director, and former director of the On-Farm Odor/Environmental Assistance program. "Low hog prices are the reason the others aren't getting done."

"If you do have odor problems and know it, an assessment may be a way to help you stay in business, especially if you want your operation to grow," says Bentley.

Not only will you benefit from the assessment, but the positive image it relays can be priceless. If your neighbors believe that you're making an effort to be an environmental steward, then they may not notice an occasional odor.

Common Environmental Challenges

Don't think that you're alone when it comes to environmental challenges. Most likely, every producer has faced or is facing some right now.

The National Pork Producers Council is feeding a computer full of data from its On-Farm Odor/Environmental assistance program. The idea is to identify producers' common environmental strengths and challenges. "By sharing information, we can maximize our resources," says Dan Uthe, National Swine Information Center director, and former OFO/EA program director.

Here's a look at some of the common odor and environmental challenges you and your fellow pork producers face.

1. Earthen manure containments or lagoons. Keeping them working with things like proper loading rates and sediment removal. Also monitoring to prevent leaking.
2. Inside buildings. Dirty ventilation systems that let odors travel on dust particles.
    Those of you using pull-plug systems don't always empty the shallow pit each week. Or if you have multiple buildings, you're pulling all of the plugs at the same time, which releases a tremendous amount of odor. Staggering that activity can help reduce odors.
3. Outside of the buildings. Standing puddles are an unnecessary odor source, especially if manure gets into these. Another problem is uncontrolled vegetation around the buildings and lagoons.
4. Lack of an appropriate manure management plan or understanding how to use it.
5. Lack of an emergency manure management action plan, especially one that is in writing.

Uthe says a complete list of the top 10 problems should be available within the next few weeks.

OPPORTUNITIES WILL ABOUND IN THE
NEW PORK INDUSTRY

Jim Carlton

But you might have to re-tool and learn new risk management and other skills to adapt to the structural changes taking place.

Despite unprecedented changes in the pork industry the last several years, many opportunities will remain for veterinarians desiring to continue to serve the industry. This was the gist of the message conveyed by University of Georgia swine veterinarian David Reeves during this year's Howard Dunne Memorial Lecture at the American Association of Swine Practitioners Annual Meeting.

"Diagnostic and technical service areas will likely expand," says Reeves. "Production veterinarian positions will evolve, and opportunity in this area will be a function of one's knowledge of biology, business skill and communication skill. Production positions will be more managerial in the future and will be even more systems oriented, he adds.

"Essential to success in production will be skill in bench marking and process management. Through each, production managers will be better able to predict and act upon change. As margins narrow, there will be less emphasis on growth and more on optimizing resource efficiency. Throughput with efficiency will be expected of management."

Reeves points out that even though in the past swine practitioners have prided themselves in being adaptive, "the recent dynamics of the swine industry have challenged many of our long-held persuasions."

Practitioner Survey

To gain perspective on how these dynamics are affecting the pork production industry, Reeves surveyed 33 swine practitioners, asking them to list the greatest concerns of producers, the greatest concerns to the industry, and the greatest concerns to swine practitioners. (See accompanying bar graphs.)

Lessons from the Past

In light of the structural changes occurring in the pork industry, is there a way to construct an effective and sustainable pork production company for the new swine industry?

Despite dramatic improvements in technology and production systems, we still don't have a manual to help in this regard, says University of Georgia swine veterinarian David Reeves.

"However, the ghosts of the last 30 years offer some very telling lessons for those willing to heed their cry," he says. These include:

- Opportunity is fleeting.
- Size, though important, isn't the only factor determining sustainable success.
- Efficiency is just as important as throughput to success.
- Financial, price and production risk must be managed with equal skill and vigor.
- The cycle of innovation and continuous improvement affects sustainability and the relative importance of each is dependent upon time and circumstance.
- Benchmarking and process management can improve the chance for success.
- Variability matters as it improves our understanding of chance occurrence.

Reeves says that the events of the last two years have demonstrated that "no one golden nugget guarantees success in the business of swine production. Like a three-legged stool, production, market and financial risk equally place production companies at risk. The importance of one or the other is dependent upon the business dynamic at the moment.

"Sustainable businesses are those that anticipate and act upon these risks with equal fervor. In other words, the business model must be comprehensive and address risk with every tool possible."

Successful production veterinarians, says Reeves, have realized the importance of risk management skills even though they may have had to acquire them after graduating from veterinary school.

These 33 practitioners included private practitioners, consultants, corporate practitioners, academicians and technical services veterinarians from the United States and Canada, he says.

The practitioners surveyed perceived that producers are most concerned about the environment, market access, industry structure, business management and risk management, food safety, animal welfare, market price, access to genetics, labor, government regulation and disease, according to Reeves. "Some described producers as fatigued and worried, particularly about their children's future, equity for retirement, public perception of the industry and perceived public anti-agricultural sentiment. Environmental regulations, market risk and financial risk are the most common concerns for producers.

Industry concerns perceived by the practitioners surveyed included the environment, competition from foreign pork, competition from other meats and protein sources, product acceptance because of pork safety and quality, industry structure, labor, business risk, government regulation, animal welfare, introduction of foreign animal disease and public perception of the pork industry, says Reeves. "A few expressed concern about the leadership of the industry and the future role of NPPC."

Future Role?

As for the concerns of swine practitioners themselves, Reeves says, "every responder expressed concern about the future role of the swine veterinarian and most expressed concern about the future of the business of swine practice. There is a consensus that as the industry continues to consolidate, there will be less opportunity in private practice.

"Many responders expressed a need to gain new skills that prepare (them) for the new swine industry, including leadership, communication and business skills. Others suggested that food safety, health and disease, limited treatment options caused by government regulation, animal welfare, government regulation and industry structure are of concern. Some expressed concern over matters such as unethical practice and time required for travel. Lack of technician support, secretarial support and control over veterinary matters within production companies also are listed."

PRODUCER CONCERNS

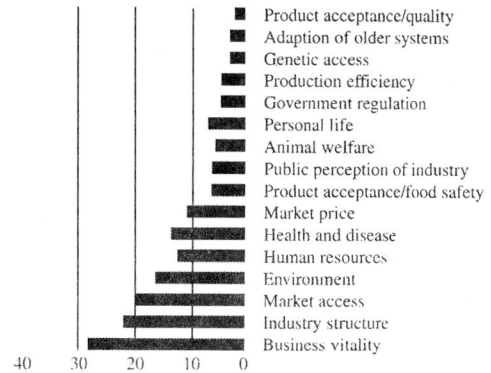

Product acceptance/quality
Adaption of older systems
Genetic access
Production efficiency
Government regulation
Personal life
Animal welfare
Public perception of industry
Product acceptance/food safety
Market price
Health and disease
Human resources
Environment
Market access
Industry structure
Business vitality

40   30   20   10   0

INDUSTRY CONCERNS

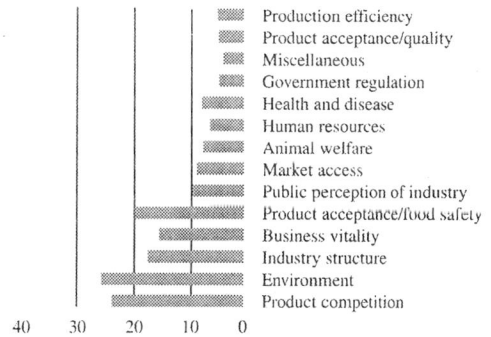

Production efficiency
Product acceptance/quality
Miscellaneous
Government regulation
Health and disease
Human resources
Animal welfare
Market access
Public perception of industry
Product acceptance/food safety
Business vitality
Industry structure
Environment
Product competition

40   30   20   10   0

PRACTITIONER CONCERNS

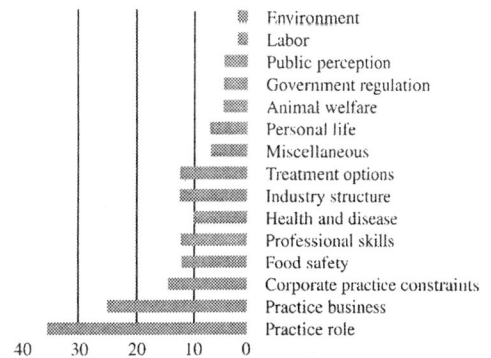

Environment
Labor
Public perception
Government regulation
Animal welfare
Personal life
Miscellaneous
Treatment options
Industry structure
Health and disease
Professional skills
Food safety
Corporate practice constraints
Practice business
Practice role

40   30   20   10   0

These graphs illustrate results of a survey of swine practitioners who were asked to rank their perceptions of the greatest concerns to pork producers, the pork industry as a whole and swine practitiones.

Reeves says that while these issues aren't necessarily new, "two years of low market prices have exacerbated these concerns. Further, they have compressed the time necessary to deal with these issues. Whether these concerns are threats or opportunities will ultimately depend upon one's strengths and weaknesses," says Reeves.

He suggests that as we look toward the future, "it is imperative that we not deny change, but rather that we embrace it. We cannot always control change. However, we can use our skill to manage change. We must understand the new dynamics and adopt new technologies which will secure our individual positions as leaders in the new swine industry. For many of us, this will require re-tooling."

How to Remain a Viable Part of the Pork Industry

Natalie Knudsen

Strategic planning and adapting to a rapidly changing client base will continue to be a key to remaining a part of swine veterinary medicine in the future.

The practice of veterinary medicine has changed more in the past five years than in the 20 previous years—and no one realizes this more than swine practitioners.

Nearly gone are the days of clients with 50 to 150 sows who relied on their practitioner to kill pigs, treat sick animals and recommend vaccination programs. Today's swine producers ask tough questions on nutrition, genetics and marketing and expect knowledgeable answers from their practitioner.

Adapting to Change

"The veterinarians who are still in practice are the ones who took the time to do some strategic planning," stresses Don Draper, DVM, PhD, MBA, College of Veterinary Medicine, Iowa State University. Draper made a presentation "Shaping Your Veterinary Future," at the Iowa Veterinary Medical Association Meeting last winter in Ames.

Those veterinarians who have remained viable in swine practice "looked at who their clients were, what products and services those clients needed, and how as a swine practitioner they could meet those needs," says Draper.

And practitioners contemplating their future role serving the pork industry need to be doing the same thing today.

Most practitioners in mixed animal practices have watched the number of swine clients decrease. "While sow numbers have remained stable in our practice, we've lost 40 to 50 percent of our producers," says Jack Creel, one of six veterinarians in the Valley Veterinary Center, Cherokee, Iowa.

Many of the traditional 80- to 130-sow farrow-to-finish herds are no longer in business due to financial (and structural) changes in the U.S. pork industry But there have been a number of these type of producers who have joined as part of a segment of a larger swine operation. "Clients wanted to remain in the swine business and asked to be part of a sow co-op so they could focus on growing/finishing pigs and crop production for their farming operation. Our role was to bring those clients with similar interests together into one operation by organizing sow co-ops," notes Creel.

Large Operations

Clients with larger operations on multiple sites with multiple owners have presented a major change for the Ottumwa Veterinary Clinic, Ottumwa, Iowa. "My job description has changed from treating sick animals to preventive medicine and management," says Steve Menke, practitioner in the mixed animal practice. "We're more involved in the big picture of the operation including cash flow, payroll, facilities, manure management, employees and environment. I think today's veterinarian is the best person to integrate all of these areas."

As practitioners at the Veterinary Medical Center, in Williamsburg, Iowa, contemplated forming and managing large swine co-ops, they did not want to ignore their traditional producers. "One consideration was 'Are we going to upset our traditional producers,'" notes Mark Brinkman. "Those producers were feeling threatened by the larger units."

Brinkman and the five other practitioners in the practice ultimately decided, "If we don't provide the services the producers want and need, there won't be a need for us." The group formed VMC Management, a swine management business. But the veterinary clinic still remains a mixed animal practice, serving pork producers, cow/calf operations, small animals and exotics.

"Today's clients demand a higher level of service stimulating a need to specialize within swine practices," remarks Paul Yeske, partner at the Swine Vet Center, P.A., a swine specific veterinary practice in St. Peter, Minn. "The more we educate them, the more challenging questions they ask of us," says Yeske, adding that increasingly those questions have moved away from animal health issues specifically to questions that integrate the financial impacts of health and production decisions.

Handling the Management

Whole unit management demands that swine practitioners become knowledgeable in diverse areas, including human resources, tax forms, genetics and nutrition.

Management is a service where swine practitioners can provide additional services to increase their income, according to Menke. "If we're involved in an operation's management, we're going to do whatever it takes to make it successful. Our reputation and credibility are on the line."

Additional responsibilities in unit management lead to specialization within a practice. Brinkman says they've learned to specialize within the practice and delegate responsibilities. "Each veterinarian has a more specific area of expertise today," he says. And additional support staff are needed as production record keeping needs increase. Lay people are utilized in swine production units under veterinary supervision to perform animal services allowing the swine practitioner to make more efficient use of their time.

"I don't view myself as an expert on everything," says Creel. "I view my role as a facilitator." Nutrition, genetics and marketing are several areas where outside expertise is often utilized in their practice, he says. "As a facilitator, I serve as a link between my producers and outside knowledge sources who can answer their questions and benefit their operation."

Successful management of employees plays an important role in large units. "If we have control of the employees through hiring and training we can control production better," says Brinkman. "Our management fee is based on the productivity of the unit." The challenge is to get employees to take pride in their work and view it as a career rather than a transitional job.

According to Creel, handling the record keeping and financial side of a unit—including paying the vet bill—has brought a new perspective to the practitioner.

To take on these new roles in the consolidated swine industry, practitioners are increasingly turning to outside education sources to keep them up-to-date on the latest innovations in the swine industry and expand their knowledge base in non-traditional areas such as statistics and financial planning.

"Veterinarians are not necessarily good businessmen because we spend so much of our time and education on disease," notes Brinkman. "That's why we need to seek out opportunities for advanced training in the business and management aspects of these operations."

Planning for the Future

Draper believes that unprecedented change for practitioners will continue into the foreseeable future. He highlights the following four areas that deserve your attention over the next five years and beyond.

1. Veterinarians will be increasingly looked to as knowledge brokers. This knowledge becomes an economic asset when transferred to your clients.
2. There will be more business function integration. Information from all production sites and segments of the operation is used to make decisions positively impacting financial health.
3. Food safety will continue to grow in importance and opportunity for practitioners. Opportunities for practitioners in pre-harvest food safety include certifying the care, handling and quality of the animals leaving a producer's unit. Food safety is an area where all the practitioners agree they see opportunities. "Health and epidemiology of how disease interact within the system will continue to drive

profitable healthy production," stresses Yeske. Veterinarians may be called upon to certify the safety of the food product as it leaves the farm. Detailed records on nutrition, genetics, drug usage, pathogen load and animal care provide the basis for this system.

4. The focus will be on syndromes of risk. Areas of risk in all producer operations must include environment, genetics and behavior considerations rather than diseases only.

Draper adds that information integration will not only play a key role in terms of knowledge but in terms of businesses within the swine industry. For example, "We're working with pharmaceutical companies on research trials," says Brinkman. "Five years ago we'd never expected to do this kind of work but now we view it as an opportunity." Record keeping plays into integration as nutrition, genetics and environmental decisions for the future are based on past animal performance.

According to Draper, information integration will also take place over the Internet. "Swine practitioners will find that the exchange of technical information over the Internet is phenomenal. It also serves as an additional means of communication with clients and the swine industry in general."

Improving production efficiency and servicing the needs of the client are the bottom line for the future. "Practitioners will continue to see less hands-on work and become more involved in information management and analyzing data," predicts Joe Connor, swine practitioner with Carthage Veterinary Service, Ltd., Carthage, Ill.

"I think a veterinary education prepares us with the critical skills of problem solving and accessing information," needed to be a part of the new pork industry, stresses Connor.

Menke and his partners strive to develop integration and management systems that will serve as prototypes for other producers to follow. "Not only are we working to bring non-corporate people together to make a new system, but we're working to expand the systems already in place." Financial stability of the producers involved is an important factor in the success of each system.

"I ultimately went to school to become a veterinarian," says Brinkman. "But times change and if you don't adapt you'll be left behind. If you're lucky enough to find a career you like, you'll adapt and learn in order to survive."

MARKET HOGS?

John Beghin and Mark Metcalfe

An international perspective on environmental regulation and competitiveness in the Hog Industry

Since 1994, the United States has been a net exporter of pork. Annual exports for 1999 and beyond will probably exceed 600,000 tons. The expected continued expansion of the pork industry has made the United States a threat to its European competitors (USDA-FAS). Traditionally, the United Sates has under exploited its comparative advantage in hog production. Despite its low feed and labor costs, heterogeneous genetic stock and small-scale production handicapped competitiveness. Recently, however, the industry achieved new scale economics in production and has taken advantage of information coordination. New, larger production units facilitate more competitive hog-processing technologies, reduce transportation costs, and systematically produce attributes valuable in the export market.

In some U.S. states, however, this new industrial organization is accompanied by a geographical concentration of production, raising environmental concerns. Even so, the challenge for producers in Northern Europe to comply with tough domestic environmental regulations may further enhance the comparative advantage of pork production in the United States (Beghin and Metcalfe).

This article assesses the extent of recent environmental regulation affecting hog industry competitiveness in the European Union (Belgium, Denmark, the Netherlands), Poland, Taiwan and 25 individual U.S. states. We emphasize the heterogeneous and evolving nature of environmental regulation, which varies dramatically from state to state and across countries. Despite the geographical disparity in regulations, there is everywhere a common trend toward introducing more stringent and new policy instruments. These policies respond to public environmental concerns and anticorporate farming sentiments, although "large farm" means something different in each country and state.

Rising Environmental Concerns and Heterogeneous Regulatory Response

Worldwide, environmental concerns linked to hog production are increasing. In the European Union (EU), especially in the Netherlands, Belgium, and Scandinavia, concerns about waste disposal are forcing producers to adopt costly waste management techniques, or to scale back their production capacity. The 1991 European Community Nitrate Directive, the central legislation regulating European water quality, prescribes that nitrate concentration in water should not exceed 50 parts per million and that nitrogen applications (after plant intake) should not exceed a standard of 170 kilograms of residual nitrogen per hectare per year. Most hog producing regions in Northern Europe are declared "vulnerable" because they do not meet these nitrogen and other standards. More drastic policies will progressively bring these regions into compliance. The new policies increasingly limit output and export expansion.

In the Netherlands, for example, manure production rights (MPRs) regulate phosphate emissions. Manure rights owned by the farm cap its phosphate production. Historical farm production (based on 1987 pig herd size) and available land determine each farm's manure quota and prevent the scale of existing farms from expanding. The government reduced the historical MPRs by 30 percent in 1995. Farms may purchase MPRs from other farms, but each transaction reduces the MPRs by 25 percent. Farms can trade their MPRs within regions and sell MPRs from manure-surplus regions to manure-deficit regions, but not vice versa. Surplus regions correspond to the traditional hog and cattle producing regions of the Netherlands. Additional restrictions on MPRs limit trade across livestock activities, further constraining hog production.

Ammonia quotas further limit hog production in some Dutch regions. Farms must not exceed a maximum concentration limit for ammonia, which varies by location. Farms in excess can buy ammonia quotas from deficit farms, but only from within the county. A well-established market facilitates ammonia quota trades.

The current direct cost of environmental regulation (waste handling and treatment, manure production rights, ammonia reduction) in the Netherlands is significant and rising. Den Ouden of Wageningen Agricultural University estimates the direct cost of regulation to be between 5 and 10 percent of the average total cost of hog production, but he estimates costs could go much higher (up to 24 percent of the average cost per hog) if new regulations prohibit nitrogen and phosphate emissions. Dutch policy makers imposed a drastic 10 percent reduction of all farm herds in 1998–'99 and plan further decreases, with a minimum of a 20 percent decrease from 1998 levels to be achieved by 2000.

Denmark, too, faces constraints, but ones less onerous. Land-use requirements (manure/land ratio) and operation permits constrain expansion of Denmark's hog production. Permits typically require concrete manure storage facilities with a one-year capacity and stipulate various setbacks from water sources. Hog farmers can spread manure in concentration that does not exceed 170 kilograms of nitrogen per hectare per year. This standard does not account for nitrogen intake by plants and hence, exceeds the EU standard. Manure spreading must meet seasonal restrictions based on the form of the manure (solid or liquid) and the types of crops to which it is applied. Manure must be directly incorporated into the land.

Danish farmers unable to meet the manure/land requirement can spread waste on farmland belonging to mineral-deficit farms. The transfer from surplus to deficit farms must be documented, but in practice it is hard to monitor. Since 1990, grain production must be followed by a cover crop to take up nitrogen. Sixty-five percent of the land on each farm must be covered in winter with a crop. As in the Netherlands, and since 1994, farmers are required to maintain nutrient balance sheets and fertilizer management plans based on animal waste and fertilizer use. The balances must be sent to the Danish Ministry of the Environment. Fines are levied on farms that produce excess nitrogen. Finally, new regulations prohibit the creation of operations larger than 15,000 head. In the long run, the Danish government plans ambitiously to reduce nitrate emissions by 100,000 tons per year, about half of Danish agricultural emissions.

While environmental regulations have significantly affected hog production in the Netherlands and Denmark, it is premature to know their impact on competitiveness in the other countries surveyed. Belgium, with its profitable and well-organized hog industry, appears the least ready of the three EU countries examined to seriously tackle its water quality problem. Nutrient standards in the Belgian Flanders allow for up to 400 Kilograms of nitrogen from all sources per hectare per year. Still, according to work by Lauwers, van Hylenbroeck, and Martens, current regulations induce an estimated cost increase of $1.92 to $5.27 per hog. Regulations planned for 2002 would increase this cost up to $17.75 per hog. For now, Belgian regulators have resisted enacting direct reductions in hog production. Since 1997, in an effort to preserve the family farm, small Belgian hog producers (800 hogs per year or less) face more lenient environmental regulation than do large-scale operations. Larger farms must process and ship their manure long distances.

The U.S. hog industry also faces regulatory pressure, but relative abundance of land mitigates some of the cost associated with waste management. Investors can build new, large hog operations in locations better able to absorb manure waste, and new operations comply with regulations at a lower cost than older established operations when the latter require retrofitting.

Many U.S. states have their own sets of environmental regulations which affect hog production. These regulations often include setbacks, approvals for facility design and waste systems, and nutrient standards. The severity of regulation varies from state to state, as does the rate of change in regulation. For example, North Carolina and South Dakota had "lax" regulation in 1994 but were among strictly regulated states by 1998. By the end of 1998, three-fifths of the U.S. states we reviewed had proposed further regulatory legislation (NACPTF).

New regulatory tools have emerged, such as bonding and moratoria on new operations. Bonding requires a farm to show financial responsibility in the event of a waste lagoon spill or closure (closing or abandonment) through insurance, security bond, letter of credit or participation in a state's waste facility closure program. In 1994, both tools were almost unheard of outside Mississippi, but by 1998, eleven states used either bonding (Illinois) or moratoria at the state level (North Carolina) or county level (Colorado).

As in the other countries, U.S. regulations discriminate between large and small operations. Colorado's Confined Animal Feeding Operations Control Regulation of 1992 distinguishes between animal feeding operations and concentrated animal feeding operations (5,000 head or more). Concentrated feeding operations cannot discharge manure into waters of the state. In Oklahoma, operations exceeding 2,000 head face stricter setback requirements.

Taiwan has limited regulation of hog waste. The 1991 National Water Pollution Control Act Amendments set emissions standards for livestock operations larger than 200 head. The standards limit biological oxygen demand, chemical oxygen demand, and suspend solids. The average cost of complying with these standards decreases with farm size and is estimated to be nearly 7 percent of total cost for operations with less than a 1,000 hog capacity, down to about 3 percent for large farms of 5,000 head or more (Taiwan Livestock Research Institute). With their rising affluence, Taiwanese consumers increasingly value water quality. New zoning regulations are emerging to limit the number of hog farms in the watershed of rivers used for drinking water (USDA-FAS).

Poland, the last country reviewed, is emerging from its transition to a market economy. Its acute environmental problems, however, do not originate in agriculture, and the link between water quality and livestock waste is currently not scrutinized. If Poland joins the EU as projected, it must eventually comply with EU regulations. The EU has explicitly acknowledged more flexible enforcement of environmental regulations for future members, such as Poland, to account for their lower income levels.

The accompanying table summarizes and ranks the importance of environmental regulations in the countries and states reviewed.

Environmental Regulation for Hog Producers Uneven among Locations But Increasing

Strict environmental regulations faced by EU producers increasingly compromise their long-run competitiveness. The United States has more land and lower manure concentration, both important environmental assets, giving the U.S. pork industry a competitive advantage over its European counterparts. Still, some states and local authorities have introduced moratoria and other regulations which curtail their pork industry.

We also noted biases in environmental regulations against large operations. The Netherlands taxes scale expansion through manure production rights. Belgium, Taiwan, and the United States grant exemptions from storage requirements, permits, and other standards to smaller producers. In the United States, Iowa, Missouri, and Nebraska grant exemptions to small farms, but so, too, do new producing states, such as Colorado. Countries use a variety of policy instruments to mitigate environmental damage from hog operations. The Netherlands grants tradable ammonia and manure rights. Some U.S. states require bonding. Commonly used instruments include setbacks and approval facility design and waste systems.

Finally, we note that environmental considerations increasingly affect trade agreements. For pork trade, EU producers would likely support an "upwardly leveled playing field" in environmental regulations for livestock production. However, negotiated international regulations would most likely not be binding for them, and the variety of policy instruments would render such negotiations tedious.

PORK

THE OTHER WHITE MEAT® CAMPAIGN DRIVING POSITIVE ATTITUDES

Farm Chronicle

More US consumers of pork in recent years have favorable opinions of pork in recent years thanks, in part, to the producer-funded Pork. The Other White Meat® advertising and promotion campaign.

The survey found that the 87 percent of Americans who have seen or heard about pork through the industry's Pork. The Other White Meat advertising, public relations, retail and foodservice efforts are much more likely to think favorably about pork in all areas, from taste to nutritional value. Eighty percent of consumers aware of the campaign also reported they would most likely be eating pork within the next month, compared to 60 percent of consumers unaware of the pork campaign.

Fifty-six percent of consumers surveyed this year have a favorable opinion of pork, up from 47 percent in 1993. Among consumers who are aware of the checkoff-funded Pork. The Other White Meat campaign, favorable opinions were almost 60 percent. Consumer attitudes and usage of pork, beef and chicken were measured using a Meat Attitude and Usage Tracker Survey

"The Pork. The Other White Meat campaign was a bold move by the US pork industry when it was launched in 1987," said Steve Schmeichel, a producer from Hurley, SD, and chair of the producer-driven Demand Enhancement Committee. "These survey results reinforce why pork producers continue to use Pork. The Other White Meat as the cornerstone for our marketing and advertising programs.

In 1999, $20.9 million, 57 percent of the national pork checkoff dollars, were invested in domestic demand enhancement programs. USDA is forecasting US pork consumption for 1999 at 53.9 pounds per person, 1.3 pounds higher than last year and the highest per capita consumption since 1981.

Pork's overall consumer favorable rating of 56 percent is up from 54 percent taken in May 1998. The survey also found that beef's favorable rating increased from 64 to 67 percent this year while chicken remained unchanged at 84 percent.

"Results of this survey will help the US pork industry target future communications efforts and narrow that gap by emphasizing pork's key area of appeal to consumers: being a lean, white meat that is something different from their usual routine," said Schmeichel.

The Meat Attitude and Usage Tracker telephone survey, funded by the pork checkoff, surveyed 1,000 consumers between the ages of 25 and 70.

EFFECTIVE INFILTRATION

Jason Gerke

Because of a cow's unique ability to convert forage resources to muscle, beef production and profitability are tied to the amount of precipitation that can be utilized by live plants.

Author Charles Dudley Warner once wrote, "Everybody talks about the weather, but nobody does anything about it."

While it's true we can't control the wind or predictably manipulate rainfall, there are management tools available that will enable you to take advantage of the precipitation your land receives to maximize production. More rainfall doesn't mean more production if that water doesn't stick around long enough to be used.

The Hydrologic Cycle

The amount of water on earth never changes. But its form and location is constantly in a changing cycle. Precipitation that reaches the ground either evaporates, infiltrates into the soil or runs off the soil surface to ponds, lakes or oceans. That small amount of precipitation that infiltrates into the soil is the key to plant life and productivity. Synonymously, plant life and soil surface characteristics affect the amount of water that infiltrates and is retained in the soil.

"Most ranching occurs in arid or semiarid regions so generally you don't have enough water to begin with. But when it does rain you want that water to go into the soil and not run off," says Butch Taylor, superintendent of the Texas A&M research station in Sonora, Texas.

The conceptual model of water use on rangeland illustrates how forages beneficial to beef production are able to utilize only a small percentage of the precipitation that falls on south Texas rangelands. Where each drop of water falling on your pasture goes depends on the climate, geology and the vegetation on your land.

Improving Rainfall Effectiveness

Whether you produce beef on arid range, lush grass pastures or the short grass prairie, management of your forage and soil resources can improve the effectiveness of rainfall by:

❏ Reducing runoff
❏ Avoiding over-grazing
❏ Controlling undesirable weeds and brush
❏ Minimizing evaporation

Reduce Runoff

The process of harnessing the sun's energy in green plants and converting that energy to protein through cattle would never be possible without sustainable soil resources. Surface runoff represents a serious loss of water and the erosive nature of runoff transports soil and nutrients from the land. The loss of soil reduces the quantity of water that can be stored at any one time. This not only limits forage production but results in plants running out of water faster, increasing the frequency and severity of drought.

"The first step in capturing rainfall is ensuring that you have as much of the ground covered as possible. That means covered by green growing plant material or by dead plant litter," says Jim Gerrish, forage specialist at the University of Missouri's Forage Systems Research Center in Linneus, Mo. "If pastures are always being grazed down to 1 1/2 to 2 inches you are going to have bare ground. If you have bare soil between plants, you can count on having run off. But if you have plant residue on the soil surface it will act as a sponge and hold the water in place."

Vegetative cover also protects the soil's surface from raindrop splash. This is important because water infiltrates into the ground through pore structures in the soil's surface. When raindrops hit bare ground they break down soil structure, closing the pores to infiltration.

Surface water sources, including ponds, lakes and rivers, depend on the runoff of rainfall to maintain water volume. However, any negative impact of water infiltration on surface-water volume is minimal. Improved plant covers can potentially increase infiltration and total run off at the same time.

"You can get more runoff even with more grass because grasses use less water than brush," explains Bob Knight, associate professor of Rangeland Ecology and Management at Texas A&M University. "Grass plants use some water in the summertime. Then when the winter rains come the amount of water in the soil builds up and it takes less to recharge the soil to peak water holding capacity. The result is more surface runoff when additional rains come in the spring."

Avoid Over-grazing

During the summer, the evapotranspiration rate of forages growing on pasture around the Missouri Forage Systems Research Center will run 1/4 inch to 3/10 of an inch per day.

"How often do we get 1 3/4 inches of rain every week? Not very often," says Mr. Gerrish. "Capturing as much spring rain as possible to store in the soil profile is what really keeps your pastures growing in the summer time."

How you manage your pastures has a significant impact on water holding capacity. In fact, poor management this winter can negatively affect infiltration by spring. Feeding hay to your cowherd on the same pasture all winter allows cattle to graze every plant down short. In addition, the hoof action during the wet part of winter and spring compacts the soil.

"Avoiding keeping cattle on the same pasture throughout the entire winter is going to make that pasture more resilient when spring comes and better able to take in water and store it," says Mr. Gerrish.

Though environments differ, the factors that allow soil to hold more water are similar. A key ingredient in increasing water-holding capacity is adequate forage residual. Repeatedly overgrazing pastures hurts soil structure, slows infiltration and reduces the amount of organic matter on the soil surface.

Grazing plants too short is a result of over stocking a pasture. And the more animal pressure you put on a pasture the more the soil gets compacted. This problem is magnified when overgrazing decreases root mass that would otherwise help break up the soil. Leaving that taller residual is going to help your root system.

"Here in the Midwest we usually get the freezing and thawing that breaks the soil compaction back down. At a low to moderate stocking rate, each winter will correct the compaction that you created the previous summer," says Mr. Gerrish. "But with sustained high stocking rates, we don't see it going back to as low a level each year."

Control Undesirable Weeds and Brush

While having any plant cover is better than bare ground, not all plants in a pasture contribute to livestock production. Weed and brush plants take water from the grass that could potentially be used for livestock production. It's estimated that mesquite use as much as 100 gallons of water for each pound of above ground plant material produced. Perennial grasses use less water, between 40 and 75 gallons, for each pound of usable plant material.

In research trials done in south Texas on mixed brush, researchers found that forage production can increase significantly upon implementing brush control measures.

"Research I have done with juniper has shown that the surface area below and around a mature tree may only be able to produce 200 to 300 pounds of forage per acre per year," says Darrell Ueckert, a Texas A&M professor of rangeland ecology and management. "But if you kill that tree and leave it standing, within two years that same land under that tree can be producing 2,000 pounds of forage per acre. That's how severe juniper competes with forage plants for water."

Clearing brush not only frees up water supplies needed by more efficient grasses and forage plants, it can increase the amount of water that percolates below the root zone replenishing deep aquifers. Dr. Knight explains that herbaceous forages only extract moisture out of the surface because their roots reach down only 3 to 5 feet. Brush roots will go down 15 or 20 feet.

Minimize Evaporation

"Cedar and other woody plant species also catch a significant amount of rainfall in their canopy that would potentially go into the soil to grow grass or even recharge the underground aquifer," says Terry Bidwell, rangeland management specialist at Oklahoma State University. "When rain water is caught in the canopy much of it will evaporate and never even reach the ground. And the more arid the climate is the more significant that interception of water is."

Interception loss refers to the amount of water that adheres to above-ground vegetation and directly evaporates back to the atmosphere. Because of the differences in structure, area and texture of plant surfaces, vegetation greatly influences interception. Forage resources can be manipulated to reduce evaporation and canopy interception while at the same time protecting soil surfaces from the effects of heat and wind.

"Juniper has a very dense canopy year round, which can intercept a lot of rainfall," says Dr. Taylor.

In Sonora, Texas, precipitation averages 22 inches per year according to Dr. Taylor. If rangeland has Juniper canopy covering over 30 percent of the surface, he estimates only about 16 inches of that rainfall will actually reach mineral soil because of interception. Allow the canopy to cover 60 percent, a common occurrence, and only about 9 inches of rainfall would ever reach the ground.

THE REALITY OF VOLATILITY

Paula Mohr

Record milk price highs and lows were the norm in the 1990s. Are you ready for the ride in 2000 and beyond?

With the good ol' days of price supports and stable milk prices fading into a distant memory, now is not the time to allow yourself to become nostalgic.

Instead, it's time to bone up on Federal Order reform and learn how your milk is priced. Note that you'll be paid more accurately for the components in your milk. And cheese prices will still drive your milk price.

Milk checks from February through May will probably pay between $10.05 and $10.80/cwt. for Class III milk (formerly the Basic Formula Price), predicts Bob Cropp, University of Wisconsin agricultural economist. He sees the all-milk price averaging $12.50 for the year, down from $13.75 for 1999.

Lower milk prices—maybe not this low, however—really shouldn't come as a surprise to producers. In reality, the dairy industry has seen its highest and lowest prices in the last decade when the support price hovered around the $10/cwt. mark and the government got out of the business of buying surplus dairy products. Now consumers dictate milk prices by buying and not buying dairy products.

Knowing that we're once again in the low part of the milk price swing, there are a few management choices you can make to ease your operation through penny-pinching times.

❑ *Protect your feed supply.* With the probability of drought this summer, farm management consultant Jim Kastanek, Albany, Minn., is encouraging his clients to line up corn for all of 2000 and soybean meal through September. "Don't get caught blind-sided by feed prices going up and down," Kastanek says. If your storage is full, buy cheap corn now and use the stored feed in June, July and August.

In parts of the Midwest, snow has been light, so some producers might be dealing with alfalfa winterkill next spring. "We're trying to get extra hay supplies lined up, too, just in case," he says.

❑ *Keep pace with inflation.* If you want to maintain your profit margin, Kastanek reminds producers that they'll have to increase production to cover inflation. "If

---

we have 3% inflation and your herd produces 20,000 lb. of milk, your goal should be 20,600 lb. and then you'd hold your own in your economic sphere," he says. Too often, he sees producers slide back and stay content there. But meanwhile, costs of production continue to climb.

❏ *Know your true cost of production.* Too often, Bob Matlick, with Moore, Stephens, Frazer and Torbet, LLP, Visalia, Calif., hears producers say they produce milk for $9/cwt. But it turns out some things were omitted from the expenditure side. When milk checks are small, it's crucial to know what it costs to operate your dairy. Matlick says know your production cost and actual break-evens on a cash-flow basis for operating expenses, cattle replacement costs, principle and interest payments and owner's draw and compensation.

Once you establish those numbers, review them monthly or at least quarterly and compare them to your projected and actual milk prices. If you see shortfalls, communicate with your lender immediately and let him or her know your financial situation.

"Don't let your lender come to you if there's a problem," Matlick says. "Be proactive. They may extend the amortization or forgo the principal payment."

He encourages clients to think through "what if" scenarios. For example, if feed costs increase 10¢/day, can you boost production to offset that?

Knowing that it takes a small community to support a dairy operation, Matlick likes to include a producer's nutritionist, veterinarian, banker and other key support people in dairy management discussions. "It helps keep us all on the same page with the same objectives." Matlick says. "The better producers who are succeeding get everyone striving together toward one goal."

❏ *Compare your production costs to reliable benchmarks.* "Profitable dairy producers understand that farm management is under their control," says veterinarian and consultant John Ferry, Adam, N.Y. "Some dairy farmers think if they can't control milk prices, everything is out of their control. That's not true. Costs of production vary from farm to farm, so farm management is under their control."

As Ferry works with his clients, he compares operations that are similar to each other to formulate the most accurate benchmarks possible. Together, they create a standard chart of accounts for management to maintain through the year. From this bookkeeping setup, they identify additional areas to benchmark, and strengths and weaknesses on the farm. Then they put together an action plan to reduce production costs.

❑ *Lock in your milk price.* It's too late to reduce your risk with getting into futures and options right now. But it's something to consider when the milk price climbs again and you want to protect yourself from the next fall.

Talk with your milk buyer, too. More cooperatives are signing contracts with their producers that lock in a milk price. Producers who ship to Land 'O Lakes had about two weeks in December to decide if they wanted to lock in a cheese price for all of 2000. Cheese buyers approached the co-op with the idea.

"We gave producers a short time to respond and they did in droves," says Land 'O Lakes spokesman Terry Nagle. Several hundred producers from across the U.S. signed contracts for up to 75% of their production.

In a nutshell, producers would get their pay price for the month with all the usual additions and deductions, then their final price would be adjusted up or down based on $1.33/lb. block cheddar cheese. If cheese prices are lower than $1.33, producers would get the additional income up to $1.33. If cheese prices are higher than $1.33, then that difference would be deducted.

From the response the co-op received, Nagle says they will offer similar contracts again.

FARMS OF ALL SIZES CAN ADAPT

Thomas Quaife

You don't have to grow in cow numbers, necessarily. But, you do need to grow in your management skills.

Brian Brown is a soft-spoken, unassuming young man from south central Wisconsin—not the kind of person you'd normally think of when describing the expansion trend that's sweeping the dairy industry. After all, expansion involves risk and a little bit of ego. Shouldn't we be talking about the high-flying entrepreneurs out west?

Well, when you scratch below the surface stereotypes—and they are just stereotypes—you will find that the expansion trend involves a cross-section of the dairy industry. Perhaps the only common denominator among these people is their desire to be a part of the industry well into the new millennium.

Brown, of Belleville, Wis., is a prime example. He recently expanded his dairy from 100 cows to 270 cows. The milking herd is no longer housed in an old stanchion barn; instead, it is treated to the comforts of a new free-stall barn. A new milking parlor provides added efficiency.

More cows mean more employees. In the past year, Brown's outside labor force has grown from a few part-time workers to two full-time and five part-time employees. This allows Brown and his wife, Yogi, some extra freedom to attend their children's soccer games or 4-H activities.

Others, like Brown, are making their own moves to remain competitive and find a better lifestyle. It's happening. It's reality. Instead of bemoaning the fact that expansion could lead to an over-supply of milk, thus driving down prices, we need to recognize the driving forces and adapt to a changing environment. For some producers, it may mean initiating expansion or modernization projects of their own; for others, it may mean a graceful exit from the dairy industry.

If you plan to expand or adapt, here are three key strategies to use.

Plan Your Steps Carefully

Brian Brown utilized consulting help from Land O'Lakes Dairy Business Services prior to his expansion. Together, he and the consultants went through all of the production and financial numbers—going back three years in the past and then projecting three years into the future. The numbers were solid enough for Brown to get bank financing.

Business planning is a must for all dairies, regardless of size. That was illustrated recently in a Farm Credit Services (FCS) project involving 37 Wisconsin dairies ranging in size from 50 cows to 2,000 cows. Following modernization, all of the dairies were able to cut their break-even production cost per hundredweight and raise their net cash income per cow by significant amounts. (See the article, "it's not the number of cows, it's how you manage them.")

Going into the study, this particular group of dairies could be termed as "cautious optimists," says Michael Krutza, president and CEO of Farm Credit Services of North Central Wisconsin. As a group, they were likely to be viewed by their neighbors as successful, yet fiscally conservative.

IT'S NOT THE NUMBER OF COWS, IT'S HOW YOU MANAGE THEM

Thirty-seven dairies in Wisconsin have made major progress by working with business consultants from Farm Credit Services of North Central Wisconsin.

The following charts show progress made by the three size groups studied—50 to 200 cows, 200 to 350 cows, and more than 350 cows. It's interesting to note that break-even production cost and net cash income per cow were pretty much the same across all three size groups following modernization.

BREAK-EVEN PRODUCTION COST ($/CWT.)

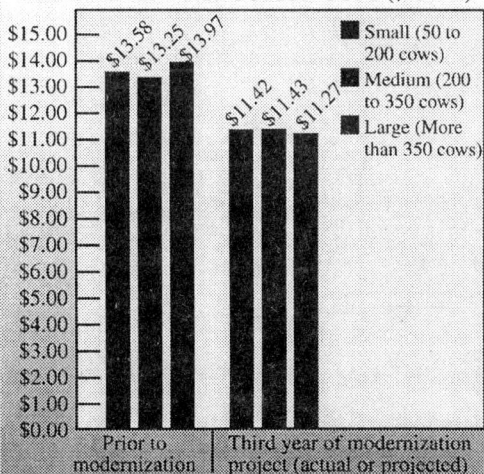

Prior to modernization: Small $13.58, Medium $13.25, Large $13.97
Third year of modernization project (actual or projected): Small $11.42, Medium $11.43, Large $11.27

Legend: Small (50 to 200 cows), Medium (200 to 350 cows), Large (More than 350 cows)

NET CASH INCOME PER COW

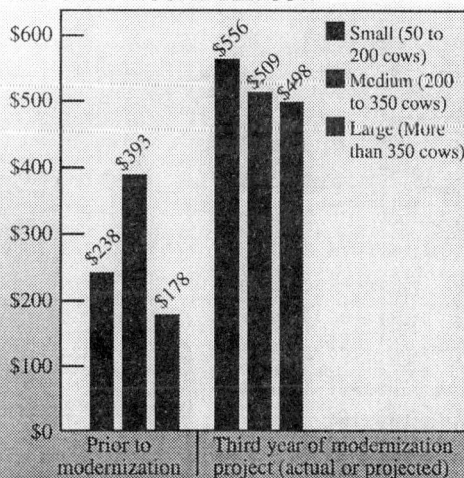

Prior to modernization: Small $238, Medium $393, Large $178
Third year of modernization project (actual or projected): Small $556, Medium $509, Large $498

Legend: Small (50 to 200 cows), Medium (200 to 350 cows), Large (More than 350 cows)

Source: "Knowing Business Factors is Key to Modernization," Farm Credit Services of North Central Wisconsin.

Each of the dairies worked with FCS business consultants in planning their modernization projects. Sixty percent of the producers in the 50- to 200-cow size range decided to retrofit their existing facilities with either flat barn parlors or in-barn pit parlors. Two of the producers stayed in stall barns. Those producers with more than 350 cows built new state-of-the-art facilities.

It was interesting to note that all size groups—50-to-200 cows, 200-to-350 cows, and 350-plus cows—ended up with pretty much the same break-even cost per hundredweight and net cash income per cow after their modernization projects. In other words, many of the important economic parameters were size-neutral.

The ability to manage one's assets properly—not sheer size—appears to be the most important influence on dairy farm profitability on a per-cow basis.

Make Wise Investments

By keeping a close eye on numbers and investing in only the most productive assets, Rich and Johanna Borba were able to grow their dairy at a phenomenal rate.

After graduating from college, Johanna Borba started out with a herd of 240 cows on a leased dairy facility near Merced, Calif. Soon afterward, she met and married Rich. Together, they expanded the herd incrementally—600 cows and beyond. The secret was to buy the most productive assets—the cows—and lease the less-productive ones, such as land and facilities. They were able to periodically re-adjust their loans, borrowing against the cows so they could add even more animals to the herd. For three and one-half years, they exercised that strategy until they were ready to buy their current dairy in Winton, Calif., which houses 1,400 cows.

All of that has taken place since 1995.

Johanna Borba admits some of it was luck, along with having a family that was already well-established in the dairy business. The young couple's ability to control— not necessarily own, but control—those assets which gave them the highest return on investment has been a key. The dairy industry offered them an opportunity, and they took advantage of it.

They chose a 1,400-cow dairy for various reasons. It was the size of dairy that Johanna's father ran—and ran quite efficiently. Their double-22 milking parlor can handle 1,400 cows per eight-hour shift, which keeps the parlor working around the clock on a 3X milking schedule. With 1,400 cows, the Borbas believe they can still have some personal knowledge of the cows, whereas they felt that a 5,000-cow dairy would be too impersonal. And, 1,400 cows is big enough that the Borbas can afford specialized employees, such as a full-time feed manager, who can concentrate on one specific area of the dairy.

"You have to constantly run the numbers," Rich Borba says. You must assess, "What would happen if I add more cows, or change this?" That was the biggest lesson that Rich Borba learned from his wife's father, Henry teVelde. Prior to marrying his wife, Rich was a partner in an independent insurance firm. So, in all reality, looking at the numbers was really not new to him.

"A lot of people think dairying is a lifestyle, or a right, but this is a business," he adds. To succeed, you must run it as one.

Minimize Your Risk

It's no secret that dairying is a potentially profitable field for people who can keep their cost of production down and margins high—and that attracts outside investors.

People get excited when they see the return on assets and return on equity that some of the better dairies have achieved in recent years, says Terry Smith, CEO of Dairy Strategies, a Madison, Wis., business consulting firm. Granted, 1998 was a pretty phenomenal year in terms of milk prices, but some dairies were able to achieve a 15 percent return on assets, he adds. "The best-in-class over the last couple of years have broken a 20 percent return."

Smith is working with two groups from outside the traditional dairy structure, both of whom are constructing 2,000-cow dairies in Iowa. One group is made up of local investors engaged in other forms of agriculture, with some external capital coming from outside the community. A crop farmer heads up the other group.

Both groups understand price volatility in the dairy industry, and that milk prices may fall into the $10-to-$11 range this spring. "But, they also see the benefits of developing closer linkages up the food chain with processors and manufacturers," Smith says. In other words, they are developing long-term milk contracts with processors which assure them of minimum prices, with the potential to realize higher prices if the market goes up. The processor shares in the price risk through futures contracts, put options and other trading instruments.

The investors were already accustomed to using risk-management tools in other agricultural industries.

Yet, few people in traditional dairy circles take advantage of the risk-management tools available to them. According to *Dairy Herd Management's* 1999 Management and Market Profile Study, 83 percent of dairy producers surveyed had not used the futures markets in the past 12 months. Of those who did use the futures markets, 68 percent had used it for grain only. The picture isn't much better with forward-contracts: 79 percent had not used them in the past 12 months. Of those who did, 78 percent had used them for grain only.

The outside investors who use risk-management strategies may know something that the rest of us don't.

You don't have to be a big investor to share in the opportunities afforded by the dairy industry. But, you may need to do some of the same things they do—albeit on a smaller scale.

The only limits that exist are the ones that you put on yourself.

LARGE DAIRIES TEND TO BE MORE PROGRESSIVE IN CERTAIN AREAS

When you survey a cross-section of the dairy industry, as *Dairy Herd Management* did recently, it becomes apparent that large dairies do certain things which give them a competitive advantage. For instance, dairies with 200 or more cows are more likely than smaller dairies to employ the services of business consultants, engage in farm enterprise cost analysis, and use futures contracts and other risk-management tools.

The following data are from the magazine's 1999 Management and Market Profile Study which included the responses of 637 producers, each with 50 or more cows.

FARM ENTERPRISE COST ANALYSIS

Do you analyze costs by the various farm enterprises?

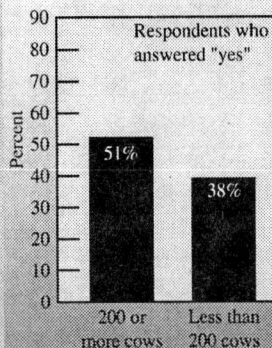

Respondents who answered "yes"

- 200 or more cows: 51%
- Less than 200 cows: 38%

Source: *Dairy Herd Management* 1999 Management and Market Profile Study

COMMODITIES FUTURES MARKETS

In the past 12 months, have you used the commodities futures markets?

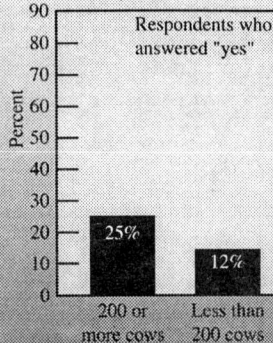

Respondents who answered "yes"

- 200 or more cows: 25%
- Less than 200 cows: 12%

Source: *Dairy Herd Management* 1999 Management and Market Profile Study

CONSULTANTS OR ADVISERS

Does this farm use paid consultants or advisers?

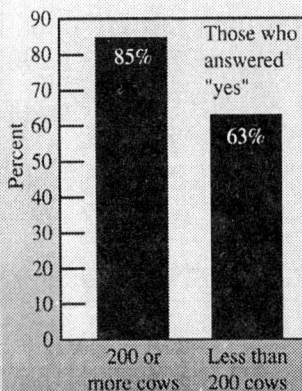

Those who answered "yes"

- 200 or more cows: 85%
- Less than 200 cows: 63%

Source: *Dairy Herd Management* 1999 Management and Market Profile Study

FORWARD CONTRACTS

In the past 12 months, have you used forward contracts?

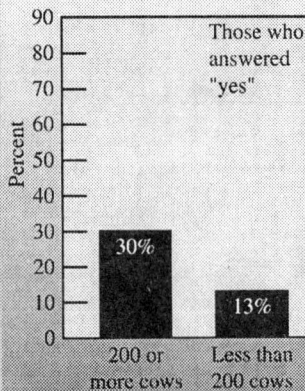

Those who answered "yes"

- 200 or more cows: 30%
- Less than 200 cows: 13%

Source: *Dairy Herd Management* 1999 Management and Market Profile Study

SPOTLIGHT ON SMALL PRODUCERS

Sharla Ishmael

If you ask Gene Sollock about his custom grazing operation on 54 acres outside of Iola, Texas, you'd better be ready to think in terms of numbers. He knows, off the top of his head, how much income each acre generated during the last 13 months—down to the penny.

He knows how many inches of rainfall his land was blessed with during each of those same months, and he can tell you the exact dates of the last two times he applied fertilizer—Jan. 14, 2000, and Jan. 13 . . . 1995! And Sollock can reel off the percentages of protein—both crude and digestible—that each of his various classes of forage had to offer just a few months ago.

His is not a large operation in terms of land area, but it is intensively managed and well-run. He neither feeds nor raises any hay, and the cattle that graze his pastures don't expect any supplemental feed, even in winter, even in the middle of a drought. Sounds impossible, but at age 73, Sollock says one needs a good challenge to look forward to.

Born and raised not more than three miles from his current place, Gene Sollock has been involved with his land one way or another since 1965. After serving his country during World War II in the Tank Corp, 33rd Infantry Division, he earned a degree in agricultural education at Sam Houston State University.

He then spent the better part of 22 years teaching high school agriculture classes in Pasadena, Texas, and later in Conroe, Texas, where he retired in 1982. After his wife, Ruth, retired from her high school counselor position, the Sollocks moved to the Iola ranch and turned a weekend endeavor into Gene's full-time challenge.

"Everything that we do here is trying to develop an animal that is suitable for the market and do it in a way that is natural," he explains. "Animals were developed to use large amounts of roughage, that's the way the good Lord made them. We try to afford them with all their nutrients through plants. What you see them grazing is what they get."

Ankle-deep Clover

It works, too. In early February, after a year that saw barely more than half of the area's "normal" rainfall, I'm standing with him in a green field, ankle-deep in clover.

From *The Cattleman, April 2000* by Sharla Ishmael. Copyright © 2000 by Texas and Southwestern Cattle Raisers Association, Inc. Reprinted by permission.

97

TABLE 1. STOCK DENSITY AND GRAZING INCOME, OCTOBER 1998–OCTOBER 1999

Month												
Oct	Nov	Dec	Jan	Feb	March	April	May	June	July	Aug	Sept	Oct
No. Head												
71	123	123	71	71	69	154	154	105	57	83	83	83
Grazing Income												
$558	$923	$953	$550	$497	$535	$1,155	$1,194	$788	$442	$643	$623	$643

$9,504 was earned from 54 acres in 13 months, which averages $18.54 per acre per month, or a total of $176.02 per acre.

Elsewhere down the road, a herd of crossbred mommas and their babies bawl like they've just been weaned. Actually, they're just looking for the feed truck because their grass gave out a while back.

While it is true that the field we are standing in hasn't been grazed in almost 90 days, that's just part of the flexibility that Sollock's operation affords him. His pasture program is based on clover and its alluring ability to fixate nitrogen in the soil, combined with about 13 years worth of intensive, rotational grazing.

"This is just old, worn-out cotton, corn and peanut land," Sollock says. "It was farmed to death and then they ran goats and cattle on it. It was really tired. Our clover acts as a soil builder, and at the same time, it does all of the fixation of the nitrogen in the soil. That way, it is able to feed your plants as they come out in the spring and carry them on through summer and into fall.

"The nitrogen is released very slowly throughout the groing period," he explains. "So, you've got a plant that is high in nutrient value and your cattle are able to do real well on it. The clover really is our building block."

Earthworms and Dung Beetles

Sollock also relies on earthworms and dung beetles to keep his soil in good shape. They break up manure and carry it down into the soil, but their presence means he has to pay extra attention to the timing and frequency of management practices like deworming and fertilization. However, he says he doesn't mind because, "They make an alleyway for the water to go down into the soil."

"To keep from depleting the soil, everything is done in almost a cyclical manner," Sollock explains. We don't take anything from the soil; we don't even cut it for hay. Everything is done by grazing and then we return it to the soil. We don't have to drag pastures. We don't have to use herbicides to control weeds. We do that with stock density.

"When you put 150 head in a 2.25-acre paddock, they will graze it down in a 24-hour period—small weeds and everything. Then, as your sod becomes thicker and better defined, you get better growth from your grass and it overpowers those weeds."

Sollock rotates the cattle through a series of 24 paddocks, with each paddock getting at least 21 to 22 days of rest between grazing. The cattle, which are mostly stockers, are moved at intervals ranging from one to four days. He waters them from a private well and has meters on each line running to the pastures, which helps him detect leaks right away.

The meters also help him keep an eye on water consumption, which gives him some idea of the relative moisture content in his forages. As well, it tells him something about where the animals are at weight-wise, by using a rule of thumb that says weather conditions aside, on average, cattle consume 1 gallon of water per 100 pounds of body weight.

"It's pretty interesting," Sollock admits. "You'll find that it is very consistent. Within a 30-minute time period, you'll have little variance in the daily water consumption if temperature is regular."

He also uses those water lines (1.5 to 2 miles worth of underground pipe) to deliver certain medicines with the help of an inductor.

"We mix it up so that each animal will get his share of the medicine (for dewormer, scours, etc.) over two or three days," he explains. "That way, I don't have to pen them. That's a little progressive for some folks. But I'd rather have the cattle out in the pasture working than in the chute stressing. Plus, I've already passed 73 years of age and I pretty much do everything here myself. I don't want to be in the chutes either."

Training to Hot Wire

Like many of the things he does on the ranch, Sollock is fairly particular about the way he handles cattle—even though somebody else owns them. For instance, he takes special care to train newly arrived steers and heifers to his electric fences before he sends them to pasture.

"That eliminates a lot of problems," he says. "Whenever cattle are brought in, we'll put them in a large pen with a hot wire on the inside of the perimeter fence. I put ribbons on it to entice them to come over and smell it and touch it. As soon as they find out it won't hurt them if they stay away from it, they won't challenge it. If you just put them into the paddocks without breaking them to hot wire, they become frightened, and they'll run through it."

While they are in the pen, he also takes time to walk among them and move them about to teach the cattle not to be afraid of him. He doesn't holler at the cattle or slap his legs to get them to move. When the cattle are taken to the paddock, all they have to do is figure out where the gates are when he wants to move them.

"When I go out there, I call them and they gather around and follow me," he says. "It's very easy to do. In 20 to 30 seconds, you can move 100 to 150 head, no problem at all. You don't have to cowboy and run the cattle and all those things. After a few weeks, you'll always have some cattle that are more docile and curious than others, and they'll start coming up to you and smelling of your hands and licking your pant leg or sleeve."

That's the kind of care that makes a custom grazing business successful. But despite his infatuation with good soil management and attention to the animals, Sollock's bottom line is still all business.

Grazing Contract Options

"What I'm doing is grazing another person's cattle, and I do the management of the grazing, the moving and all," Sollock explains. "You have so many options in this game. This last set of cattle I had on a grazing contract where I was grazing them so much per head, per day. The next time, I might do it on a weight-gain basis. But then it's necessary for everything to be weighed in and weighed out and there are a number of other things to consider.

"I guess I use a lot of this for continuing education for myself. It kind of becomes a challenge because people say you can't make money in livestock—that's not right! You can. I realize what I'm doing here is on a small scale, but all you do is make your paddocks larger, get more cattle in and do it. The results are the same or better."

Sollock's interest in rotational grazing and soil management piqued in the early '80s when he attended conferences in Mississippi hosted by a publication called the Stockman Grass Farmer. Visiting with the folks he met there who practiced intensive grazing and seeing what was happening to his place as he divided it up for rotational grazing convinced him he was doing the right thing.

"I could see that there was improvement being made with the grasses and the soil," he says. "We have just continued to refine it. I'm in a transition period right now, trying to decide if I want to graze my clover or leave it alone to develop a seed crop. If I was a little more assured that we were going to have more rainfall, I would go ahead and start grazing.

"I would graze it on up until the first or middle of April, then pull the cattle out and let it make a seed crop. We can harvest, normally, around 70 to 100 pounds of seed an acre. The seed is running $4 to $5 per pound. We split that income with the young man who does the combining, but it still works out to an additional $125 to $140 an acre. Then you can start grazing again about the middle to last of June," Sollock says.

Best Money You Can Spend

Soil tests and plant analyses are two key tools that Sollock relies on to coax as much productivity as possible out of his land, and at the same time, minimize expenses. He's only fertilized twice in the last five years, but that's just because he only does it when yearly soil tests show a deficiency.

"As far as plant analysis, I think that's something that should be used more than we do," he admits. "It probably would be beneficial for us to check our growth and plant availability maybe two or three times a year.

"If I have standing grass out there in late October, everybody tells me it's no good. The biggest part of it's not," Sollock says. "But this land's had clover growing over it for a number of years, so the nutrients have built up and the plant has more in it to start with. So, you do a plant analysis and find out—hey, this stuff's still 12 percent. You may find out it's only 4 percent if it's been grazed down. But you do a plant analysis and then you know. That is probably some of the best money that can be spent."

Sollock believes this type of management shows its true value during a drought. Otherwise, instead of being able to run an average of 80 to 90 head a month on his 54 acres, he says he'd be hard-pressed to run 30 or 40 head.

"People go by the place and think I'm irrigating," he smiles. "But I don't have anything to irrigate with. I am dependent on rainfall. The thing is, when I get an inch of rain on the place, I hold it there. We've got the sod and the organic matter in the soil and it will hold the water. I get full use of most all the rain that falls, because it doesn't run off."

According to his records (see Table 1), this strategy helped Sollock generate an average of $13.54 per acre per month during the dry months of October 1998 through October 1999—a grand total just shy of $10,000.

"The thing is, I'm not having to invest any additional money into it," he explains. "I just provide the supervision over the cattle grazing and soil management. I realize that I am retired and have other income coming in, but I don't want to do it for free or lose money. I heard somebody say the other day that the smaller operator can't compete. Well, I can compete with the best of them," he smiles.

THE ANCON OR OTTER SHEEP

Dr. D. E. Salmon

The following article is a most graphic example of how breeders can select for an attribute that looks worthwhile but, in fact, destroys the usefulness of an animal.

To the best of our knowledge there are no pictures of the Ancon sheep.

The Editor

Seth Wright, who owned a small farm on the banks of Charles River, about 16 miles from Boston, kept a little flock of common sheep composed of one ram and 15 ewes. In the year 1791, one of the ewes produced a male lamb of singular appearance, differing, for no assignable reason, from its parents by a disproportionately long body and short bandy legs, whence it was unable to emulate its relatives in their scaling the neighbors' fences and luxuriating in forbidden pastures. The neighboring farmers recognizing this excellent characteristic of the new sheep, advised Mr. Wright to kill his old ram and reserve the younger one for breeding, which advice was followed. The first season, two lambs only were weaned in his likeness. In the following years, a number more, distinguished by the same peculiarities. Hence proceeded a strongly marked variety in this species of animals, before unknown in the world. It was called by the name of the Otter breed. This name was given from a real or imaginary resemblance to that animal, in the shortness of the legs and length of the back, supposed by some to have been caused by an unnatural intercourse; by others, perhaps as fancifully, from fright during gestation. It is only certain that otters were then sometimes seen on the banks of this river. They have since disappeared.

Dr. Shattuck, who in 1811 dissected one of these sheep for the purpose of ascertaining the properties and qualities which distinguished them from the common kind, says that the sheep weighed, just before it was killed, 45 pounds. The most obvious difference in its skeleton from the skeleton of the common sheep, so far as a superficial observation extended, consisted in the greater looseness of the articulations, the diminished size of the bones; but more especially in the crookedness of its forelegs, which caused them to appear like elbows. Dr. Shattuck called them *ancon,* from the Greek word which signifies elbows. On dissecting the sheep he could not forbear noticing the comparatively flabby condition of the subscapularic muscles, which would partially account for the great feebleness of the animal and its consequent quietude in pasture. The inequality of form seemed to be confirmed in the blood.

Source: *The Sheperd, Vol. 45, No. 3, March 2000.*

Experiments in crossing changed the strain, or, to allow the expression, amalgamated the qualities of those of other breeds, so as to produce a mixed or mongrel race in too few instances to form an exception to the theory. Col. David Humphreys, of Connecticut, who reported this singular animal to the Philosophical Society of England, said that when both parents were of the Otter or Ancon breed, the descendants inherited their peculiar appearance and proportions of form, and he had heard of but one questionable case of a contrary nature. The small number of cases where the young were said to partake in part, but not altogether, the characteristics of their breed, would not invalidate the general conclusions established on experience in breeding from a male and female of distinct kind. When an Ancon ewe was impregnated by a common ram, the increase resembled wholly either the ewe or the ram. The increase of a common ewe, impregnated by an Ancon ram, followed entirely the one or the other, without blending any of the distinguishing and essential peculiarities of both. The most obvious difference between the young of this and other breeds consisted in the shortness of the legs of the former; which, combined with debility or defect of organization, often made them cripples in mature age. Frequent instances occurred in which common ewes had twins by Ancon rams, one of which exhibited the complete marks and features of the ewe, the other of the ram. The contrast was rendered singularly striking when one short-legged and one long-legged lamb, produced at a birth, were seen sucking the dam at the same time. The Ancons kept together, separating themselves from the rest of the flock when put into enclosures with other sheep. The Ancon lambs were less capable of standing up to suck without assistance, when first yeaned, than others. Here, then, was a remarkable and well-established instance, not only of a very distinct race being established, but of that race breeding true at once, and showing no mixed forms, even when crossed with another breed. By taking care to select Ancons of both sexes to breed from, it became easy to establish this well-marked race, and there is every reason to believe it could have been prolonged, had circumstances demanded it.

Although the Ancons arrived somewhat later at maturity they were said to live as long as those of our common flocks, unless in some cases where by reason of their debility and decrepitude their health was impaired and their lives shortened. To whatever cause it may be attributed, whether arising from defect in vertebrae, muscle, joint, or limb, it is certain that they could neither run nor jump like other sheep. They were more infirm in their organic construction, as well as more awkward in their gait, having their forelegs always crooked, and their feet turned inward as they walked. It was also observed that the rams were commonly more deformed than the ewes.

This breed was looked upon as a valuable acquisition from the fact that they were less able than others to get over fences. In New England, beyond which they rarely migrated, there were few commons, no hedges, no shepherds, and no dogs whose business it was to watch flocks. The small farms were enclosed by wooden and stone fences. These were generally too low to prevent active sheep from breaking out of pastures into meadows or grounds under cultivation. Crops were injured and farmers discouraged. Hopes were entertained that this would be remedied, and these hopes were partially realized.

On the other hand, the drovers complained of the great difficulties of driving these cripples to market, and the butcher that the carcass was smaller and less salable than the common sheep. It was commonly not so fat. It did not fatten so readily, owing, probably, to less facility in moving about for gathering food. In taste it differed but little from other mutton.

There was much variation in the fleece, not exceeding in quality and quantity that of the common sheep, that from a cross of an Ancon ewe and Merino ram being very silky and of the general quality of a quarter-blood Merino. Daniel Holbrook, Derby, Connecticut, made some experiments with this sheep and the Merino which are of interest and must be quoted entire. His statement was given to the public in October, 1805.

> In the year 1800 I purchased a pair of sheep called the Otter breed. This breed of sheep are well known by some, but I presume are unheard of by many others. They generally have long, round bodies, thick necks and breasts, broad hips, very short legs that stand wide apart and some of them bend outwards. They can not run or leap fences as well as others, and mine have about the same quantity of wool as the other kind, and some finer. My lambs by those rams with other sheep have generally been either of the Otter or common kind, but in some instances they partake partly of the shape of both, and I think these ameliorate the breed. In October, 1802, I obtained one of the Spanish Merino rams imported by Col. Humphreys, and put him with part of my sheep, and by this means in the spring of 1803 had some of the half-blooded lambs. Soon after these lambs had come, I put them and their dams with my other sheep and lambs and kept them together through the summer, and in the fall separated these lambs with my others from the old sheep, and to keep them through the winter. In the summer they were manifestly different, and they wintered much better than my other lambs that lay with them, and at shearing yielded one-fifth more in weight of wool on an average than my other sheep, and the quality far superior. The wool was spun in my own family. It was carded and made into cloth at Col. Humphreys' mill, and was pronounced equal to the best English broadcloth at from $6 to $6.50 per yard.

The Ancons were widely disseminated in New England at the beginning of the present century and their numbers large, but on the introduction of the Merino they rapidly declined and were represented in 1876 by a small flock of 8 in Rhode Island. It had been perpetuated through many generations during a period of eighty-five years.

The perpetuation of this Ancon variety by the hand of man is one of the facts adduced by Darwin to show that man can by selection cause great variation in animals under domestication; can mold an accidental variety into a permanent one, in fact. From this and other similar facts he lays down a proposition which should never be lost sight of by the intelligent breeder and upon the due observance and application of, which much of his success depends:

Although man does not cause variability and can not even prevent it, he can select, preserve, and accumulate the variations given to him by the hand of nature almost in any way which he chooses; and thus he can certainly produce a great result. Selection may be followed either methodically and intentionally, or unconsciously and unintentionally. Man may select and preserve each successive variation, with the distinct intention of improving and altering a breed, in accordance with a preconceived idea; and by thus adding up variations, often so slight as to be imperceptible by an uneducated eye, he has effected wonderful changes and improvements. It can, also, be clearly shown that man, without any intention or thought of improving the breed, by preserving in each successive generation the individuals which he prizes most, and by destroying the worthless individuals, slowly, though surely, induces great changes. As the will of man thus comes into play, we can understand how it is that domestic races of animals and cultivated races of plants often exhibit an abnormal character, as compared with natural species, for they have been modified not for their own benefit, but for that of men.*

*"The variations of Animals and Plants under Domestication." Charles Darwin

RAISING SHEEP IN PARADISE

Susan Schoenian

Trinidad and Tobago are sister islands in the Caribbean, the southern most in the West Indies, situated just 7 miles off the coast of Venezuela and 11 degrees north of the equator. Famous for carnivals, calypso music and steel bands, T & T, as the islands are often called, gained independence from England in 1962. Remnants of the colonial period still remain, especially in Tobago, which was once dominated by a handful of sugar plantations.

Trinidad is the larger, more populous island, approximately the size of Delaware. Tobago, 20 miles off its northeast coast—a short 15 minutes flight from Port of Spain—is barely 20 miles long and 10 miles wide. There are significant differences between the two islands, in almost every respect. Trinidad is industrialized, appearing more prosperous than other Caribbean islands I have visited, due in large part to petroleum—oil and natural gas. Food processing is another important economic activity. The main agricultural products are sugar, cocoa, citrus, rice and poultry. The country imports 75% of its food stuffs.

The population of Trinidad is quite diverse in terms of ethnicity and religion. Approximately 40 percent of the population is East Indian, immigrants brought in to work the sugar fields. This statistic can be confirmed by the large number of Hindu temples that can be seen throughout the country. Another 40 percent is Black African, descendants of the 18th and 19th century slave trade. They are mostly Christian in faith; Roman Catholic is the major denomination. Seventh Day Adventist seemed popular in Tobago. You will see an occasional mosque, as a growing part of T&T's population is Moslem. There are lesser minorities of Chinese, white and mixed ancestry. English is the major language spoken. Cricket is the national sport. English common law prevails.

While it attracts some ecotourists, mostly from Germany and England (few from the United States), Tobago remains essentially unspoiled, an tiny island paradise with picturesque mountains, beautiful beaches and unspoiled rainforest. It is home to more species of birds and butterflies than any other Caribbean island.

Tobago's population is more homogeneous than its sister island, consisting mostly of peoples of Black African descent.

Scarborough is the largest city. The island boasts all major conveniences, but is void of the commercialism found in other Caribbean island nations.

Agriculture remains an important part of the economy. Tobago is similar to Jamaica, in that sheep can be found grazing "everywhere," whereas, in Jamaica goats are ubiquitous. Livestock do not roam free in Trinidad.

Hair Sheep

A variety of hair sheep breeds can be found in Trinidad and Tobago, many of which I had not seen before. As in other island countries, the popular breed is the Barbados Blackbelly, the "antelope-like" sheep, brown in color, with black points and a black underbelly. Jokingly, we used to refer to Blackbellies as "reproductive skeletons" because while they are one of the most reproductively efficient sheep in the world, they grow slowly and appear almost devoid of muscle. However, this could not be said of the blackbellies in T & T. Many ewes showed good size and significant udder development. On the better sheep farms, lambs reached market weights in excess of 90 pounds.

Another common breed in T&T is the West African, a reddish brown sheep, with more muscle expression than the Blackbelly. Producers commonly cross the West African with the Blackbelly to improve the quality of lambs. The West African was an impressive-looking hair sheep, in my opinion. Too bad we don't have any in the U.S., nor easy access.

The Persian Blackhead is another new breed I saw, quite numerous in Tobago. It is a fat-tailed hair sheep from Africa that makes up half of the Dorper, a hair breed that is gaining rapid Popularity in the U.S. (Dorset Horn is the other breed that comprises the Dorper.) Persians are white with a black head and neck. They show good frame, but are not as well muscled as the Dorper, Katahdin or West African.

The St. Croix or "Virgin Island White" is another common breed in the islands. They are distinguished by their white color. They are heavier muscled than the Blackbellies and considered to be better milk producers.

It is only the second time I had ever seen a Wiltshire Horn, a hair sheep native to the British Isles. Someone had brought a Wiltshire horn ram to last year's National Hair Sheep symposium in Timonium, MD, but I was more impressed with the animals I saw in Trinidad. It is worth noting that the Wiltshire Horn was crossed into the U.S. Katahdin to improve size and scale. It is the only horned sheep in Trinidad and Tobago. We saw Katahdins in the two countries due to the efforts of the Maryland Department of Agriculture to export hair sheep to the Caribbean.

Production Methods

Confinement, early weaning and concentrate feeding of lambs, and accelerated lambing are typical characteristics of sheep production systems in the two countries. Most sheep and goats are kept on elevated floors with either metal or wooden slats. Others are kept on dirt.

Confinement facilities may be shed size or as large as chicken or hog houses. Feeders are usually fence line, arranged on the outsides of the pens. Waterers, too are attached to the outside, so there is little contamination with urine and fecal matter.

A salt/mineral block is often dangled by a rope, a simple idea that I had never thought of before. Everyone has cisterns for collecting water. The rainy season runs from June to December. Flooding is not uncommon.

While some grazing occurs, most feed is brought to the sheep in the form of green chop or concentrates. Some silage is fed. Small producers will cut forages by hand, whereas bigger operations will mechanically harvest the tall-growing grasses. While tropical forage grow very rapidly, they are generally not as high in energy and protein as temperate species. Protein sources are not readily available. Nutrition is further compromised by the fact that forages are generally cut and or fed when they are too mature and stemmy, and their nutritive value and palatability has declined substantially. Of course, the same can be said about forages in the U.S. where continuous grazing causes pastures to be grazed when the grass is too mature and hays are often cut when plants have already gone to seed.

It is expensive to feed concentrates in these countries. Corn is very expensive to import. Concentrate rations are mostly home-grown, composed of various by-products of food processing. Pregnancy toxemia is a re-occurring problem, caused by inadequate intake of energy and undoubtably by the lack of exercise in confinement housing.

Reproduction Methods

Reproduction in the ewe is affected by photo period—length of day and night. Sheep are called short day breeders because in temperate regions, ewes generally come into heat as day length becomes shorter. However, in Trinidad and Tobago, day length does not vary significantly, thus ewes will express estrus throughout the year.

Breeding can be continuous. The better sheep farms remove the rams and breed for three lamb crops in two years. If proper nutrition is provided, ewes are usually quite prolific, especially the Blackbellies.

Lambing every eight months requires early weaning, typically before 90 days. All lambs seemed to be creep fed. Creep panels are used to keep ewes out of the creep area—either vertical or horizontal slats. In the first creep area I saw lambs had to crawl under a horizontal board to enter the creep area. After weaning, lambs are fed for rapid gain under feedlot conditions. Ram lambs are generally not castrated.

Health Concerns

The most common problem affecting sheep and goats in Trinidad and Tobago are internal parasites, primarily the barber pole worm. Keeping animals in confinement helps keep parasites in check.

Coccidiosis, on the other hand, is more of a problem in confinement. Animals are routinely treated for coccidia, but unfortunately products such as Bovatec, Rumensin and Deccox are not readily available to put in the feed or mineral for prevention. As previously mentioned, pregnancy toxemia is a major concern in sheep flocks. Other disease problems are similar to those in the U.S.

Marketing Challenges

Marketing can be a real challenge in the islands, especially in the more remote Tobago. We visited a large sheep farm (400 ewes and growing) that had a contract to supply a supermarket. In Tobago, lambs are sold to local villagers for festivals or to buyers coming from the bigger island. Farmers in Tobago get most of their feed and supplies from Trinidad.

As in other Caribbean countries I have visited, there seems to be a great deal of optimism about the future of small ruminant production as these small countries strive to supply a greater portion of the lamb and goat that is consumed on their soil.

ANIMAL RIGHTS

Richard W. Wilcke

A new national organization promoting Thoroughbred racing, modern high-tech methods of delivering the action to consumers, and public corporations investing in major racing facilities on both coasts, are exciting and positive developments for the sport.

But over the last two decades, there also has emerged a cautionary development: namely, an increasingly widespread uncertainty in Australia, Europe and North America about the proper relationship of humans and animals. That this confusion is being exploited and exacerbated by groups openly opposed to all ownership and use of animals renders it potentially dangerous.

Clearly, it is difficult for many trainers, who consider racing a natural activity of horses and who keep their runners in peak condition, to take seriously the political threat from animal activism. While most acknowledge a fringe element, they fear nothing worse than an occasional demonstration, say on Derby Day.

There are two distinct camps of animal-care activists. One takes the traditional humane approach called "animal welfare." While admittedly these people can be a little "touchy-feely," they nonetheless advocate methods that many trainers would endorse and most could accept. At the end of the ideological spectrum, however, is found the more radical approach referred to as "animal rights." These advocates contend that animals are morally and legally equivalent to human beings (and vice versa).

On the agenda of these extremists are not merely the banning of whips and the control of race-day medications. They are deeply committed to altering the opinions of society, no matter how long it takes, and their goal is to eliminate horse sports altogether. As remote as that may seem, the record of past movements is not comforting.

Most political historians would agree, for example, that it was not the silent majority but a determined and clamorous minority that got prohibition passed in 1920. Those who had openly scoffed at the idea that the government would ever be brazen enough to shut down distilleries, breweries, saloons and liquor stores—inconveniencing, even harming, thousands of employees and customers —were in shock.

But the campaign against liquor began in earnest 25 years prior' with the founding of the Anti-Saloon League. While this organization was supported by many who merely wanted saloons out of their own neighborhoods, its leaders never disguised the fact that their own ultimate goal was to end all uses of liquor. Years later, the wording of the prohibition amendment was drafted by the Anti-Saloon League staff.

There are, today, more than three dozen national organizations advocating animal rights. Some concentrate on vegetarianism, some on "factory" agriculture, and others on the use of animals in laboratory research. The most well known of the general organizations is People for the Ethical Treatment of Animals (PETA), which was established in 1980, and whose annual budget is well into the millions.

PETA also has millions of members, many of whom probably do not advocate laws to outlaw the use of equine sports. But the leaders of PETA, like those leaders of the Anti-Saloon League, make no secret of their perspective, which is that such domestic animals as horses are no more than slaves, and must be freed.

Industry leaders focused only on promotion should recognize that racing, to survive into the 21st century, does not need the patronage of half the U.S. population. Americans have countless alternatives for leisure time. If Thoroughbred racing were popular with only a relatively small minority of them, the revenue would still be more than sufficient to support a large and prosperous industry.

But what this sport desperately requires from a majority of Americans is tacit permission to train and race horses. If it ever were to develop that the consensus opinion in this country were opposed—or perceived so—to the use of animals in sport, the number of racing fans and pari-mutuel bettors would be irrelevant.

The question is, what do these people mean when they say that animals have rights? Everyone can understand the philosophy that humans have a moral obligation to treat animals with dignity and benevolence, but what about rights?

In a philosophical sense, true rights impose obligations. The right of one person not to be killed, for example, must be offset by the obligation not to kill any other person. In other words, to respect the same right of others. Animals, by nature, are incapable of understanding such an ethic. Precisely because of their lack of cognition (or ability to reason and communicate), they assume no obligations and respect no rights.

This idea of animal rights cannot be dismissed because of two trends in Western social and political thought in the last half-century. The first is the growing compulsion to use government to perfect every aspect of the world. The second is the dangerous conceit that truth should be discerned by voting. Politically speaking, polls and ballot boxes alone can determine truth vs. fiction, right vs. wrong.

These trends are relevant because there also is no longer a clear appreciation of the basic human rights delineated by the nation's founders. Lacking any social consensus on these fundamental rights, society has seen a proliferation of attempts to attach legal claims to values and preferences, and to call them "rights."

The inevitable result has been constant political conflicts among interest groups. Consider so-called "welfare rights," which obligate some people to provide things that other people want; e.g., food, education, health care, social security and subsidies.

Through rhetorical advocacy, politicians can appear progressive and compassionate by championing such causes. But are they really fundamental rights?

By definition, they fail the first philosophical test, universality; i.e., they are not the same for everyone. As stated, every right implies fair obligation. If each individual has a right to life and property (the view of the founders), this imposes an obligation not to interfere with every other individual's exercise of those rights. They are universal because they can be extended equally to each member of society.

But a political claim masquerading as a philosophical right cannot be extended equally. How can one person or group be said to have the right to education or health care unless the obligation to provide, or pay for, that food or health care is imposed upon others? By definition, if rights and the obligations are not the same for every citizen, a legal claim cannot be a fundamental right.

What all this has to do with racing is that animal-rights activists, due to their sympathetic feelings toward animals, hold the opinion that no human should be allowed to own or use any animal, regardless of the purpose of such use or the level of care provided the animal. This feeling is only a personal value or preference of these people, one that a free society would say they are entitled to hold.

The problem is that, like advocates of other bogus "rights," they are not willing to limit themselves to persuasion. They also want to use the authority of the law in their behalf, in this case, to prohibit all traditional uses of animals, whether for the production of food, for medical research, or for sports and entertainment. They argue that animals ought to be entitled to the same rights as humans.

In truth, even most humans in the world are not secure in their rights to life, liberty and property, nor have humans ever been throughout history. Countless wars, not only that of the American Revolution, have been fought in efforts to attain them. And the intellectual debate continues throughout the world, including within the United States, to gain and to protect these principles. Rights never come easily.

Political efforts to legalize all kinds of political "rights," including those of squirrels and horses, simply muddy the water and prove how little understood is any moral philosophy. Needless to say, animals are not involved in these debates. Animals never fought a war, wrote a book, or petitioned a legislature to secure their rights.

The political groups advocating animal rights, by exaggerating the frequency and severity of animal abuse, raise huge operating budgets from donors who feel affectionate toward animals. Having no concept of rights, and little awareness of how most animals are really treated, they support the legal claims of narrow interests.

In the name of protecting animals, these organizations will continue to suppress the human right to own and use animals for any purpose, including racing. If they ever can convince the media and the politicians that their perspective is the consensus, our game will be over. Truth, in our society, is what voters say it is.

To be certain, more understanding of rights would be useful. But another response would be some means by which the industry could convince disinterested Americans that those involved in racing are so concerned about the welfare of their horses, and so intent upon eliminating all instances of abuse or cruelty, that prohibition of the sport would never be necessary. This is the strategy of the breweries; i.e., to become so visibly concerned about responsible drinking that critics are overwhelmed.

If something along these lines does not develop, the sport of Thoroughbred racing will remain vulnerable in spite of marketing successes.

SNAKE OIL

Gary West

I recently spent an hour talking horse racing with a group of people who knew little about the sport. Try it sometimes; it's enlightening.

These people never had heard of Behrens or Silverbulletday or Prado. In other words, they were average folks. Only a few of them had been to a racetrack more than once or twice in their lives. They weren't opposed to gambling. Some of them said they bet on football, or played the lottery or went to casinos, but nobody in the group wagered with any regularity on the outcome of horse races.

And I asked them why. Why were they reluctant to visit the local racetrack? Why weren't they horse-players? Why weren't they racing fans? Their answers were, I thought, rather predictable. They didn't understand horse racing, they said, and they didn't trust it.

The second explanation is much more disturbing than the first, for mistrust is much more difficult to overcome than misunderstanding. Horse racing, I believe, has a serious credibility problem, and it exists even among the sport's ardent fans—just listen someday to the denizens of the grandstand. Such skepticism and cynicism might seem unfair, especially in these debased times, when felons participate in the NBA and NFL, but it's the perception. And the perception is the reality. To overcome it, horse racing quite simply must do much better than it's doing at present.

Horse racing's credibility rests, for the most part, in the hands of racing officials and horsemen. They're the only people who can dispel the mistrust. But in some cases, they're adding to the problem.

I speak with horsemen regularly, and most of them I find to be honest and candid. Many of them I respect, and some of them I genuinely admire. And then there are the liars and deceivers. Usually they're people who embrace the anachronistic assumption that the sport is their playground and the fans a necessary inconvenience. That attitude, of course, has proven to be deleterious and is one of the main reasons horse racing no longer occupies the position it once held as the second most popular sport in America. Some of their deceptions might sound familiar:

"He's going to make his next start in the Hutcheson." No, he's actually going to make his next start in a two-turn allowance race.

"He came out of the race good." No, he came out of the race with a chipped knee.

"He's fine; he's going to run in the Florida Derby." No, he's not fine; he's not going to race again this year.

"He's going to ship to the Fair Grounds Tuesday for the New Orleans Handicap." No, he's going to be entered Wednesday in an allowance race at Gulfstream.

"She was scratched because she spiked a fever." No, she was scratched because, in violation of the rules, she received an injection.

And so it goes. Yes, horsemen can change their minds. That's easy to do, especially when, as in some cases, there's so little to change. Plans also can change. But horsemen frequently lie about their plans and their horses; and whenever they lie to the press, they lie also to the public and the bettors, the very people whose money supports all this running in circles. People deceived don't remain loyal fans for long.

Here's some advice for horsemen: Speak truthfully and candidly, or don't speak at all. Every lie, every deception, adds to the credibility problem.

Stewards also contribute, sometimes innocently. Stewards are no longer the authorities they once were. Perhaps it began in 1990, when the California Horse Racing Board overturned the stewards' disqualification of Tight Spot in the Del Mar Derby, but more and more, it seems, stewards' decisions are subject to appeal and dispute.

Moreover, their expertise in viewing races seems inconsistent, at best. I've seen horses drift several paths, bump and impede, and I've seen jockeys use their whips to strike other horses, all without any disqualification. And I've seen horses drift perhaps two feet or only brush another horse and then be disqualified. The bettors have witnessed such occurrences, too, and that's another reason they're skeptical.

In this year's June 4 Acorn Stakes at Belmont Park, Three Ring fought back in deep stretch to win by a head over Better Than Honour, who had rallied three-wide to gain the lead in midstretch. According to the *Daily Racing Form* chart, Three Ring "came again under left-handed whipping and prevailed while forcing out Better Than Honour in the final yards." And Better Than Honour "battled gamely despite being carried out by the winner in deep stretch."

Richard Migliore, who rode Better Than Honour, claimed foul against jockey Jerry Bailey and Three Ring. The stewards disallowed the objection. And many observers who thought Three Ring should have been disqualified openly wondered if the stewards allowed themselves to be swayed because the winner was owned by Barry K. Schwartz, a member of the New York Racing Association's Board of Trustees. Perception is reality.

As a rule, professional sports require officials to demonstrate their expertise in a school or training program. Major League Baseball's umpires, for example, must all attend an umpiring school. PGA officials must all be certified, and to maintain certification they must participate in a continuing education program.

The University of Louisville and the University of Arizona offer schools for racing officials. The National Accreditation Program for stewards and judges includes 60 hours in the classroom and hands-on experience at a race-track. To maintain accreditation, 16 hours of continuing education is required every two years.

Racing Commissioners International set as a goal 100 percent accreditation of stewards by 1996. But about 33 percent of today's stewards are not accredited, according to Wendy Davis, the associate coordinator of the Racetrack Industry Program at the University of Arizona. In fact, the three Gulfstream Park stewards who will work the sport's championship event, the Breeders' Cup, are not accredited.

Horse racing these days is attempting more than ever to attract new fans. But some of the attempts overlook a basic truth of business: People will buy your product only if they trust both you and what you're selling.

USE YOUR HEAD

Moira C. Harris

I have mixed feelings about baseball caps. Oh, I like them well enough. I don't wear them too often: mainly when I'm out at a show, or when I want to keep the sun out of my eyes while working around the barn. Where you won't find me wearing one—even this nifty one with my favorite magazine's logo emblazoned on it—is in the saddle.

I've always been an advocate of riding helmets, and perhaps that's because when I was a kid, I was never given the option of not donning one. It was "wear this ugly white bowl on your head or you can't put your foot in the stirrup." And after the many falls I took as a youngster, I couldn't argue that excellent logic.

As I got older, there was a lot of pressure to look cool. It seemed like the ultimate freedom: Hair blowing in the breeze, me and my friends cantering around on our sedate hunters. But luckily there was still common sense lurking in the back of my mind that said, "Hey, you're just not infallible enough to beat the odds." So I kept my hardhat on.

In this issue, we've included a story that came across our desks some months ago about a little girl's battle with those very odds. Now, as a journalist, I admit that I tend to be a little jaded about personal stories, but this one has brought me to tears each time I've read it. And it reinforces for me that I make the right choice every time I get on a horse—my horse, any horse, Western or English.

But the word here is *choice*. We do not have laws or statutes that say equestrians must wear a helmet while on the back of a horse. When I was at Equitana last year, a woman approached me with some very harsh words for featuring non-safety helmeted riders in the magazine, saying "I'll never subscribe to it. Every picture should show riders wearing helmets." That's true. Every picture should. But that's not realistic, nor feasible. Many trainers—western and English—choose to believe that their own "good riding" will keep them from harm. It's a competitor's choice to switch from helmet to huntcap instead when in the ring. It's a western/dressage/saddleseat/hunt-seat/anyseat rider's option to school in a baseball cap. But it's their choice, and to exclude these photos from the magazine is to give an inaccurate picture of the riding community as a whole, and it would mean that there would be very few pictures in the issue.

I'm not a preacher—I cannot pound the pulpit of safety and get non-helmet wearers to repent. I am not a teacher—I am unable to give the equestrian public a failing grade for not protecting their heads. I'm certainly not anyone's mom—I can't ground any kid who forgets their hardhat. I'm a communicator, and my job is to inform people about certain topics—important topics—that occasionally mean life or death. I'm also an adult rider, and I set an example that hopefully the kids *and* adults will follow. Be intelligent, become informed, read this story, have a good cry and make *your* choice.

BREEDING FOR SUCCESS

Faye Ahneman-Rudsenske

Recent discoveries in human chromosome mapping provide the potential of eliminating undesirable traits while retaining the wanted ones, and while the equine aspect of it may be a few years down the road, when available, it will forever change the face of horse breeding. The possibilities seem endless, but is mankind ready for perfection? Yet, even with man's intervention can it really even be achieved? Comprehensive genetic research has resulted in a variety of things that were deemed nearly impossible just a few short years ago, so future technological advances leave the door wide open. However, until such a time materializes, horse breeders must rely on books, educational literature, seminars and practical experience to guide them through selection procedures. The amazing thing about it is that regardless of current day knowledge, expertise and the advice of experts, Mother Nature still has the final say.

Well-known and respected breeders were asked a variety of questions ranging from the influence of distant relatives in a pedigree to high heritability rates of various disciplines. Obviously, this article is certainly not intended as the final word on breeding, but rather to provide food for thought as well as stimulate interest and conversation. Controversial or conflicting viewpoints are presented with the idea that readers may glean whatever information fits their situation. Evaluating a horse's natural potential to perform or produce are an important and essential aspect of any breeding program, and it is with this in mind that the advice of knowledgeable breeders and horsemen who have generously agreed to share their thoughts and ideas on the subject are presented.

Once again, we went out into the field and asked the experts; the breeders who have, over the years, been successful in their endeavor to breed their ideal or as close to it as possible. We begin with Bazy Tankersley of Al-marah Arabians, Tucson, Ariz. Mrs. "T," as she is affectionately nicknamed, has produced generations of Arabians that have excelled in a variety of disciplines, and her stock can be found in countless pedigrees around the world. The longevity of her breeding program and valuable insight not only provided the basis for many other breeders to get started, but saved them valuable time by reaping the rewards of her efforts.

Ancestral Influence

According to research, distance relatives exert little influence over actual genetic makeup (except in the case of linebreeding), even though either sire or dam may carry known specific traits from an ancestor. Tankersley agrees. "Very little *overall* [influence] beyond the third generation," she says. "While certain characteristics, particularly if they are homozygous, can 'pop out' several generations removed, the overall horse is not greatly influenced by any one horse back in its pedigree."

"It depends on how distant the relatives are in a pedigree," says Judith Forbis of Ansata Egyptian Stud in Mena, Ark., in reply to the question. "It is indeed possible that some sneaky gene from the past will surprise you with something wonderful—or something you wish had stayed buried," she adds wryly. "A pedigree is an accumulation and, therefore, while most breeders consider the grandparents and great grandparents to have the greatest influence, the whole is the sum of the parts and a particularly dominant mare or stallion can be seen for generations."

More than 40 years ago, Don and Judith Forbis started their own dynasty of champions. Quite literally, reams of material have been written about them and their Ansata Egyptian Stud. A leading world authority on the Egyptian Arabian, Forbis has published several books on the subject. The Forbis' breeding program began with the importation of the world famous *Ansata Ibn Halima and two other Egyptian mares. Today, the blood of their foundation stallion runs in thousands of American pedigrees.

Tom Miller of Miller Arabians, Red Bluff, Calif., also agrees and feels that after three or four generations, the genes are too diluted to have much of an affect. For 45 years Tom and his wife, Arlene, along with their son Judd, have been building a dynasty of their own. They are the oldest known Arabian breeding program in existence today that breeds exclusively for the working western division. They have produced phenomenal horses that are not only known on the open circuit, but have been used to start other breeding programs across the country. At the very heart of their breeding program is Xenophonn (Bolero x Farviews Karmal). Better known as "Zee," the 14.2 hand bay stallion retired in 1982 after winning three national cutting championships and three reserve national championships. He is also the sire of more than 20 national champions and reserve national champions. The Miller broodmare band consists of national champions and dams of national champions, and the resultant offspring, consistently and with great predictability, dominates the working western division.

"It's a ladder," states Roxann Hart of Rohara Arabians, Orange Lake, Fla., who has owned horses her entire life and been in the business since 1968. The success of her breeding program is undisputed in the Arabian. "You have to understand the traits of those distant relatives to know if that phenotype or the physical horse in front of you exemplifies the same traits," she adds. "Certainly your closer precedent ancestors exert a greater influence as their genetic traits are not as diminished. However, you also have to realize that even full siblings can have 50 percent different genetic makeup. It really depends on what traits have carried through the preceding generations."

The "Golden Cross" Theory

What about certain nicks or the so-called "Golden Crosses" that seems to have the ability to consistently produce high quality individuals from two unrelated bloodlines? Is this just a fluke of nature or smart breeding practices? Some breeders feel that it may simply be a case of one parent being strong (homozygous) in an area that the other parent is weak in and vice versa. Others think it may be an increase in heterozygosity that causes some undesirable traits to be masked and superior traits to be revealed. Whatever it is, there seems to be no real way to foretell that it will occur prior to breeding and seeing several consistent results.

Tankersley believes the homozygous theory. "Certain 'nicks' probably work because one set of characteristics are homozygous in one parent, while another complementary set are homozygous in the other parent, so each parent is apt to contribute what you want from it than what would otherwise be the case," Tankersley answers. "This was apparently the case with Indraff and *Count Dorsaz. The Blunt Crabbet horses from which Indraff sprang tended to be balanced, deep-bodied and very muscular and with quite classic heads. This was true of Indraff. *Count Dorsaz was a Wentworth Crabbet horse, much finer in bone, somewhat plainer in the head, but with a much better shoulder and much better action. More often than not, the blend of these two lines gave us horses that were freer moving than the Blunt line and more muscular and better balanced than the Wentworth line."

The Miller's most successful "nick" has probably been with Speed Princess (Ferana x Teyma Miller) and Zee. A national champion herself, she and Zee have produced six offspring that have been national champions or reserve national champions, a track record that Miller doesn't think has been equaled. "We have a theory of breeding proven individuals," he says. "It's no guarantee, but it certainly increases the odds."

"Certain nicks happen," explains Forbis. "Sometimes they are the result of planning and sometimes they happen by chance. One strives for creating a 'nick' that is popular. The *Ansata Ibn Halima/*Morafic combination in the Egyptian imports of the 60s was formidable and carries on through their descendants today. Another nick was Nazeer on Sheikh El Arab daughters, which doubled the Mansour blood (e.g., *Ansata Ibn Halima, Aswan). The Egyptian-Spanish cross (often called the Golden Cross) was exemplified by Shaker El Masri and Estopa matings (El Shaklan and relatives); Aswan nicked with the Russian mares, as witnessed by the early Russian imports. There are also numerous successful American nicks."

At Rohara Arabians, Hart finds that the answers begin with education. "You need to educate yourself, read a lot and find what has worked for other people because you are looking at a very small sampling and trying to replicate which precedents have been effective," she says. "You also need to retain mares long enough to evaluate offspring and siblings. I use combinations over and over again like my mare Emenee (*Aramus x Diamondita) and Rohara Tsultress (Ivanhoe Tsultan x Emenee)—lines that I have worked with for a long time. I had the benefit of going back to *Aramus which was *Naborr, Negatiw, Naseem and Skowronek, and you see those lines following through. Even then, some full siblings looked more like their sire Ivanhoe Tsultan. Although this was more in the past, that was a 'nick' that worked for me. I capitalized on that and used a replication of crosses."

Inbreeding or Linebreeding?

Another often-controversial subject is inbreeding (the mating of close relatives continuously) or linebreeding (matings between close or distantly related horses, but not on a continuous basis), and many breeders display mixed emotions on both sides of the issue. Whether you agree or disagree, careful analysis of your breeding program's goals are vital. Education in regards to both the advantages and disadvantages can make the difference between success and failure.

Inbreeding refers to an increase in the number of homozygous gene pairs found in the resultant offspring, which means that both detrimental and desirable traits can be expressed. The purpose of inbreeding is to fix desired traits consistently, although it has also been known to limit the horse's ability to resist disease, produce smaller individuals, physical deformities and a decreased fertility rate with an increase in abortion and stillbirth. On the other hand, inbreeding can also result in consistently passing on highly desired traits meaning that it increases homozygosity and enhances prepotency, thereby "fixing" those traits and producing a "true-breeder." However, because it can also "fix" certain undesirable traits, culling inferior animals over many generations is very important when close inbreeding is utilized in a breeding program.

"Inbreeding and linebreeding can 'fix' the best as well as the worst traits," Forbis warns. "Therefore, it is important to use these tools judiciously, bearing in mind the strong dominant traits—both good and bad—in proposed matings. I have used inbreeding (father to daughter, half-brother to half-sister—i.e., same mothers) and full brother to full sister. Some results have been better than others, but the best were worth the trials. Sometimes the next generation will prove the value rather than the direct offspring."

"Inbreeding is beneficial in direct ratio to the quality of the stock that you use," advises Tankersley. "You are much more likely to retain and pass on desirable characteristics. I linebreed extremely close. For instance, my stallion AM Double Dream (Dreamazon x CF Gai Fantasia) is not only the product of a full brother and sister mating (Dreamazon bred to his own sister), but Dreamazon was heavily line bred so the inbreeding co-efficient [which measures the number of homozygous gene pairs inherited from his related sire and dam] on Double Dream is extremely high. The disadvantage is that particular line of stallions tends to be shy breeders and this is the case with Double Dream. However, the foals he does have are extremely uniform and very similar to Dreamazon. When we linebreed that closely, we generally outcross and, happily, in the next generation we have lost the shy breeding characteristics and maintain most of the others."

Miller discovered that various outcrossing in his breeding program did not meet his standard or expectations so he too reverted to inbreeding and linebreeding. "We have two national champion Zee sons out of different mare lines and we crossed those on Zee daughters," he explains. "There's an old saying about inbreeding and line-breeding," chuckles Miller. "If you do it, it's considered inbreeding. If I do it, it's line breeding. We just keep the tail female line pretty well separated and have found that works the best for us."

"Inbreeding or linebreeding will bring about a magnification of imperfection or perfection, so, first and foremost, before you even try it, your individuals must exemplify strong traits of good conformation," Roxann Hart adds. "I have done it with the Bey Shah line, and it worked for me. I bred Bey Shah daughters to Bey Shah sons three times and got three champions."

Balancing Less Desirable Traits

Often times while attempting to fix desirable traits, other unwanted traits seem to accompany them. Risk is an inherent part of any breeding program, so how can breeders eliminate the unwanted and yet retain the wanted traits?

Miller is blunt. "A lot of people ask me how I cure this habit or that habit and I tell them I don't worry about. I just eliminate it out of the breeding program. If the horse has a hole, we don't fool around as we don't want to reproduce that."

"All matings carry some traits that are less desirable than others," Tankersley says. "It is important to try and identify which traits are dominant and which are recessive. If it is a recessive trait, you must try to weed it out by culling. If it is a dominant trait, it is very easy to weed out because it manifests itself. For instance, we had one stallion that threw sickle hocks. I did not understand the above theory at that time, and I eliminated the stallion and most of his progeny. This was a mistake because I learned later that his sickle hocks were dominant and that I could safely have kept most of his progeny that did not have sickle hocks as they would not carry that gene."

"I think you are always dealing with one or the other, but you can improve on the traits that you want to discard," says Roxann Hart. "For example, many times a desirable trait is a very long and vertical neck. Sometimes the neck can get so long that the back can be too long as well. While keeping that neck is desirable, you may have to use a shorter, more compact cross to shorten the back. Balance is essential and that is the goal to always keep in mind."

Culling

"It's a fact of life that you often have to give up something to get something," Forbis says logically. "I choose what is most important to me and strive to come as close to the 'ideal' as possible by determining what I can and can't live with. We have always practiced rigid culling of colts by gelding them and then selling or providing them to adults or youth who will use and appreciate them," Forbis says. "We have no qualms about eliminating mares that cannot be 'bred up' or have a serious fault that seemingly cannot be bred out."

"Culling is extremely important," Tankersley says. "The hardest thing is when you have either a beautiful individual or one that has a been a great performance horse, but then produces below level progeny. The temptation is to keep trying. No matter how good the horse is as an individual, if a mare has three substandard foals by different stallions, out she goes. From the beginning, a majority of our stallions' foals have to be satisfactory or above. In our breeding program, it only takes one season to prove a stallion because we have seen foals out of this line for so many years that we can evaluate them quickly, and we don't fool around in proving the stallion. We breed him to enough mares his first season so that we can pretty well tell. Our herd is large enough for us to afford using several of our mares, and we also stand unproven stallions at a very low fee which enables us to see foals from a variety of mares."

"To be successful, you must always improve," Hart responds. "You always hope for a foal crop that is more level so culling is an economic as well as a genetic factor. I've been breeding horses since 1968, and as your selection process gets more refined, you should expect to have a very large quantity of higher quality individuals. A lot of mistakes can happen if a decision is made too early. Some individuals mature slower than others do, so there is no 'set' age to make that determination. For initial success levels, it takes two foal crops before a stallion is 'proven.' Of course, for a sustained true progenitor, I think you need to see the grandget."

Miller is strict in his breeding program criteria as well. "Each mare has to prove herself *before* she goes into the broodmare pen, and then they have to be easy breeders, easy foals, good mothers and produce something on a national level in order to *stay* in the broodmare pen," he emphasizes. "I guess our culling is pretty rigid," he admits, "but breeding is an ongoing process and it takes a long time to see the results. Sometimes things can jump up out of the clear blue. While it may be a freak of nature, you need to figure the percentages over the long haul to prove it."

Breeding Top Individuals

In general, Tankersley agrees with the rule of thumb that only the top 40 or 50 percent of all mares and less than five to 10 percent of all colts are of sufficient quality to maintain as breeding stock. "However," she points out, "in a herd such as ours where we had the opportunity to cull, select and linebreed for 50 years, I would say that 75 percent of our mares are top quality, but still not more than 10 percent of our colts."

"I believe you should only be breeding the finest and providing your evaluations are correct, I think those percentages would probably fall," Hart says. "Breeding in and of itself is risky business. It's not something money can buy. You need luck and sometimes a gut reaction as well."

"I think that percentage is too high," Miller argues. "You can't raise a race horse out of a mule. Although there are no guarantees, your odds are better if both sire and dam are physically and mentally capable of doing the job. I think a lot of people just want to raise a foal and increase the horse population. They don't necessarily breed for quality."

"I can't speak for other bloodlines," Forbis interjects, "and while 10 percent may be true of the stallions, I believe that more than 50 percent of all well-bred Egyptian mares are of sufficient quality to be maintained as breeding stock, especially if breeders practice very selective matings. One foal crop can give a very good idea if you know your bloodlines and have experience; three or four should definitely tell you. We have been able to determine the quality of our stallions' production with the first foal crop because our stallions are selectively bred for prepotency."

Grading

As far as grading (the outbreeding of a superior stallion to below average mares in an effort to raise the quality of the mare's offspring), Tankersley does not grade up Al-Marah mares. "We try to breed the best quality horses that we possibly can," she says. "This is not going to happen if the mares are not every bit as good as the stallions. However, we do encourage people whose budgets will not permit them to buy a high quality broodmare to breed to our inbred stallions in an effort to upgrade their stock."

"If the mare is of extreme poor quality and does not have a production record, I would probably not want to breed the mare," Hart agrees. "But you have to be careful and look at the mare's production record," she also warns. "I have one particular mare that is a mega-aristocrat and to look at her you would not feel that she is of significant value as an individual. Yet, she has national champions to her credit. If possible, try to evaluate the mare's offspring before you make your decision."

"We don't do a lot of grading for the simple reason that I think it takes a good individual on both sides," Miller adds. "In most cases, a stallion can't do enough to warrant breeding a poor quality mare. At best, you'll only get an average foal."

"Grading depends on what one considers 'below average,'" Forbis suggests. "There are many 'below average' mares that are extremely well bred and can, with judicious selection of a stallion, produce offspring superior to themselves. The converse is also true. Excellent mares bred to unsympatical stallions can produce average or below average individuals. The old saying, 'Blood will tell' must be considered when determining what is 'below average.'"

Beauty versus Function

They say beauty is in the eye of the beholder and sometimes man selects beauty over function, but is it possible to have both? "Absolutely," say most of the interviewees.

"I don't consider a difference between beauty and function because I think functional conformation is beauty—balance, an appropriate neck, way of going, etc.," Tankersley says firmly. "The exception to this is the head, and I see no reason to compromise as we should be able to breed functionally superior horses with lovely heads."

"A goal as a breeder is to have both," Hart emphasizes. "Last year at the U.S. Nationals we certainly succeeded with the Rohara-bred mare Rohara Reflection (Afire Bey V x Fire Serenity). She was the only purebred mare to win in both halter and performance. This year at the U.S. Nationals, another horse that I bred, Midnight Sun H (JK Amadeus x Lady Moonlight Lady SF) was Top Ten in both the Gelding Futurity and the Western Pleasure Futurity. I was also the breeding consultant on another JK Amadeus son, Wolfgang K (x *Aphrodite), who was the National Champion Futurity Gelding and Top Ten in the Western Pleasure Futurity. To my knowledge, these two horses are the only ones that crossed over in both halter and performance at this year's nationals. At two consecutive Nationals, Rohara has had horses that have won in both categories. Although they say it can't be done, why should we make a breed apart if we're breeding to the standard? Arabian halter horses should be able to perform."

"Everyone likes to ride a pretty horse," Miller alleges. "I breed for eye appeal, athletic ability and a trainable mind. If a horse doesn't have those qualities, you might as well get two pieces of *clay* and mold something that you do like. Of the three, a trainable mind might he the most important. Sometimes if a horse has the mental attitude they can overcome shortcomings in other areas, but I definitely think you *can* have all three." As an example, he points to Eleanor Hamilton's Hesa Zee (Xenophonn x Somthing Special), who will be featured next month in *Breeding for Success: Part II* as an example of an attractive horse with a great mind and athletic prowess.

"Balance and harmony *are* part of beauty and function," Forbis acknowledges. "*Ansata Ibn Halima was a classic beauty, yet also won regional park championships. There are Thoroughbreds who were handsome yet had major faults—Assault won the Triple Crown with a clubfoot. While form relates to function and good form is requisite to be outwardly beautiful, one must take into consideration the heart and spirit which often overcomes the greatest odds in achieving high performance. Even a perfect halter horse without spirit and character may stand behind a lesser individual who has 'star quality.'"

"Function follows form, but as mentioned earlier, form can be compensated for by heart, spirit and a desire to win. Certain 'form' provides a better chance to be a race horse or a park horse or a western horse, but mental capacity often determines the final ability despite perfect or imperfect form "

Environmental Effects on Heredity

Does environment play a pivotal role on heredity? Some say yes for without proper management, nutrition and training and, to some extent, conditioning that horse cannot meet his full genetic potential. All or part of these factors may contribute to a variation in appearance in horses of similar genotype and phenotype, Therefore, environmental variation must be considered when selecting breeding individuals. While qualitative traits such as coat and eye color are not affected by the environment, quantitative traits such as growth rate, size, behavior, fertility and soundness may be adversely affected. Thus, it is important to know the difference between pool management and genotype. Is the horse small because he was stunted by poor nutrition, or is it due to generations of small ancestors in his pedigree? Is his poor temperament a result of bad training/handling or is it due to his genes? is he unsound because of an inherited trait or an injury? On the flip side of the coin, is the horse genotypically superior or is it a result of outstanding care and treatment?

Although Tankersley feels that environment has little if no influence on heredity now that there is no longer a factor of survival of the fittest, she does think that it plays a role in the development of the horse from birth on.

Hart adheres to the theory that environment (through proper management) has a direct influence on heredity. So, how does Rohara maximize management and environmental influence? "Proper management helps in the future reproduction of the horse or the success of that individual," she says pointing out how the Thoroughbred industry has sought the limestone base (which is functional for bone growth) in the few places available in the world—Australia, France, Kentucky and Ocala, Fla.

"I think that you have to bring out the good traits of a horse to some degree and nourish those parts," she continues. "That also comes through in training. Letting a horse be a horse is also important, whether it's a stallion, gelding or mare. Don't closet a stallion. You want him to be a horse. You want him to have out time. We keep our stallions where they can touch other horses or geldings on either side. That's very important to horses, even in peer groups, or where they are not with an alpha horse or subjected to too much dominance. You need to nourish their temperaments even when they are outside. That is the overall management factor that can produce a better individual with a better mind. When our show horses return from showing, they go outside and it makes them more enjoyable to those around them, whether it's other horses or people. So many times in training barns, horses are just taken out of a dark stall, put on a walker, worked for a few minutes and then returned to the stall. One of the most important factors is maintaining the mind of a horse, which is one of the reasons our horses are able to return to the show ring year after year to compete. They enjoy it because they are allowed to be horses."

Miller, too, advocates letting horses be horses. Their horses are used for ranch work as well as showing and in his words, "We don't have any prima donnas here," he says. "They all have to earn their way and make a living," which reflects a growing trend for show horses. "It makes a lot of difference how a horse is raised," Miller continues. "Our theory is to turn our horses out a lot. We have plenty of room, and we let them use the entire ranch. It's mighty important for horses to know how to go up and down hills, through the brush and water—the whole nine yards. If they're raised in a 12′ by 12′ stall or a little pen, they have a lot of obstacles to overcome later in life."

Forbis points out an interesting example in regards to the influence of environment on heredity. "To some degree, nature requires adaptation for survival. I have found that while Egyptian Arabian horses bred in America maintain the classic characteristics of their breeding, they become 'dryer' and their coats shinier than when they return to their native desert environment of Arabian, the Gulf States and Egypt," she relates. "They do not, however, acquire this change by moving to the deserts of Arizona."

The Color Link Factor

So, is color, a highly heritable trait, and one of the easiest to predict and get important? Are there "preferred" colors and if so, are these colors linked with particular traits?

"I think color is a subject that is often passed over because everyone wants to discuss the feet, the neck or the tail carriage," says Hart. "In 1859, Charles Darwin stated that 'color and constitutional peculiarities go together.' I also thought that in Lady Wentworth's book she had noticed a color link, meaning that you could get certain traits based on the color of a horse. She noted that some chestnut horses from particular lines trotted in more of an English style than her greys. On that note, I certainly noticed that when I bred Ivanhoe Tsultan (*Czortan x Hillcrests Bint Tsatyr) and saw so many of his offspring, that his grey offspring tended to look different than their bay or chestnut counterparts. For example, with his chestnut offspring, I got a different ear placement. Certain types of personalities went along with certain colors. It can vary from lineage in a particular line to another. So I definitely think there is a color link factor."

Miller, too, throws out a provoking idea in regards to color. Herd sire and Triple National Champion Xenophonn has thrown mostly bays, with occasional greys when bred to grey, but never a chestnut. Coincidence or not, according to Bedouin legend, a bay is a sign of strength and stamina, two traits that definitely go hand-in-hand with Miller's stock horses!

"Color should not be a factor," Forbis declares, "but some people have preferences for color as do some judges. While personal preferences may enter into a breeding program, they should not influence a judge who should be impartial to color and partial to what is the best Arabian. The American show ring is overrun with bays— perhaps an overflow Rissalix blood which we obtained through three stallions and three mares frequently had a beneficial affect," says Tankersley. "From a not very scientific observation, I would say the majority of good cutting horses are Crabbet stock."

As the only known breeding program that breeds specifically for western working horses, Miller is proud, but nevertheless modest about Zee. "He's the Doc Bar of the Arabian cutting horse industry," Miller admits. While Miller holds that Zee doesn't have much cutting in his pedigree, although he too carries Crabbet breeding, he does have the associated traits so vital for working western horses. By Bolero, whose sire Witez II was rescued during WW II by General Patton, Zee's tail female lines goes to the Kellogg's (now Cal-Poly) great old stock horse Farana (*Nasik x *Farasin), with Skowronek, Indraff, *Raffles and Fadl scattered throughout. On his topside, Zee also goes back to Farana's sire, *Nasik, and dam, *Farasin, through the bay mare Narasa, as well as another line to Skowronek through *Raseyn. Zee and Speed Princess also share Farana as a common ancestor, as she goes back to Farana on the topside.

"In the early years, Farana was unbeatable in the stock horse division," Miller reveals. "They were called stock horse classes then, not reining, but it's the same thing. Of course, you can get a freak of nature and somebody could come up with a surefire good stock horse, but in my opinion, Zee is probably the highest percentage sire in the industry that consistently and persistently produces cutting horses."

Open Cutting Competition

How well does the Arabian stack up in open cutting competition against other breeds that have been breeding discipline exclusive cutting horses for years?

"We have proven that Arabians can successfully compete with other breeds in cutting," says Tankersley. "The only advantage the Arabian may have is their increased stamina—if they have a rigorous cutting schedule. A great majority of modern cutting horses do not have the conformation, particularly the hindquarters necessary for this work."

Although Quarter Horses dominate the open circuit, Miller-bred Arabians are well respected and Tom Miller's son, Judd, is also a triple A National Cutting Horse Association (NCHA) judge. "Some people are open minded and some narrow minded," Miller shrugs, unperturbed. "You may run into some discrimination, but it more or less depends on the individual horse. We're fortunate. We probably don't encounter as much prejudice as most people, because our horses look and work more like the type of the open [Quarter] horses. Arabians can successfully compete on the open circuit just as well as other breeds. There's more of a difference *within* a breed than there is *between* a breed," Miller refutes. "Case in point, there's less difference between a working Arabian and a working Quarter Horse than there is between a running Quarter Horse and one bred for halter. Our Arabians are similar in temperament, intelligence and trainability to the well-known cutting and reining Doc Bar Quarter Horse breeding. A lot of times at open shows our Arabians are even mistaken for the Doc Bar-bred Quarter Horses. However, regardless of the breed, it's more of the individual's ability and capability."

Halter

How consistently and which bloodlines produce halter horses, and what qualities and traits are more highly desired than others are?

"All bloodlines have certain families or 'nicks' that produce better than average," Forbis says. "Much depends on selection by the breeder. When one looks at the American show ring today, there are such a variety of combinations of bloodlines and such an inconsistency in what judges consider classic Arabian (standard) type, that this question is difficult to specifically answer. One finds that a particularly dominant stallion will be 'in' and influence a generation in this country more than 'specific' bloodlines. However, the use of Egyptian blood within the first three or four generations has been very influential (Russia with Aswan and Nil; Poland with Aswan and Nil derivatives and recently other Egyptian stallions; Spanish with Shaker el Masri and now Ansata Sinan).

"I do not avoid breeding for any specific traits; I concentrate on those which I believe are the ideal standard. However, bad dispositions would not be tolerated and characteristics which inhibit harmonious conformation are bred away from, especially underslung low-set necks (no matter how long), straight shoulders and weak hindquarters.

"In regards to size [of the horse], I say size in what? An old veteran of the racetrack once told me that if there was anything to size, an elephant could outrun a jackrabbit.

There is a standard for the Arabian horse and while there are some exceptions, the standard should be adhered to as close as possible, especially with regard to type.

"Temperament is paramount as nobody wants to live with a mean or stupid horse. Conformation goes without saying. There are few perfect legs, that's why judges are reluctant to give '20s' for legs under the European system and usually stay around the 15–17 marks out of 20. Feet uphold the horse, and there are tendencies in certain families (too small feet, ill shaped, clubby, etc.) which must be watched closely so that they are not intensified. Everything is important to make up the ideal complete horse. Here again, some things may need to be sacrificed to get others and the usage of the horse should be taken into consideration in making such determinations.

"Balance (harmony) and type go hand-in-hand. Nothing is reliable or consistent unless it is selectively and continuously bred for. Witness the diversity of 'Arabian type' in the show ring, many of which do not look like purebreds, but more like the Arabian derivatives (National Show Horses, Half-Arabians, etc.) that permeate the Arabian shows today. I believe that the straight Egyptian Arabian has collectively remained closer to the 'archetype' of what is considered the classic Arabian by artists and breeders of yore. This is obvious by attending the Egyptian Event where such extremes existing in the general Arabian shows are rarely found."

For years some of the top horses in the industry have either been bred or shown by Rohara Arabians. What bloodlines does Hart utilize in her breeding program to produce better than average halter horses? "Bey Shah," she affirms. "They have so much type and charisma. Other bloodlines that produce type include the Aswan lines and Padrons Psyche. They have been top leaders as far as the production of halter horses. For example, this year I crossed the Psyche lines on Afire Bey V and Justafire for both halter and performance prospects. This 'nick' has worked for other people, and I'm trying to achieve a good combination or harmony of both."

Hart is adamant about the characteristics her horses must possess and the direction that some breeders are going. "One of the things that people tend to overlook in the breed today is feet," she says. "That may sound like a very small part of the horse, but 'no foot, no horse.' We are seeing more and more feet and legs, particularly angulation from the ankle, from the pastern into the foot and clubfeet. This is an area that breeders really need to concentrate on in their breeding program. Of course, everything else goes with it, but this is one area that is often overlooked."

If she could only choose one or two traits, what would they be? "Providing that he has generally good conformation as a basis, I would require a horse to have a very long, upright neck that breaks well back from the poll," she pauses and adds, "and type. For a halter horse, you must have a neck, and if you are going to step up and go into the performance arena, whether it's western, English or driving, you have to have a neck in order to position the horse in the bridle."

Hart also advocates temperament and trainability. "It's 100 percent important," she says firmly. "This may be an altruistic type of thing, but I feel some bloodlines produce 'heart.' By that, I've seen many horses that have the potential and the talent, but they basically don't want to do it. There are some lines that produce this trait consistently. When you're going around the ring for the tenth time at an extended trot, you need a horse that wants to do it, wants to please and wants to stay with you. Personality is a vital component and bloodlines that produce trainable and tractable horses lend themselves more for the show ring today."

Breeding Goals

The value of any breeding program depends upon the breeder's goals and selection of the appropriate horses with the desired characteristics and traits. These factors all require careful analysis and consideration. Selection is not a concept to be taken lightly if success is to be imminent. Breeding for success requires diligence, patience and time. It is a multi-faceted undertaking with far reaching consequences and should produce a horse that will have a lasting effect on the equine population not only now, but in the future as well.

CAMP COUNTDOWN

Sarah Christie

Sending your child to summer camp can be an enriching experience for the both of you. Learn how to choose the right camp and send your kid off without a hitch.

The smell of pine needles and Calamine lotion still conjures up images of starry nights, songs around the campfire, and my very first ride on a black and white pinto pony with a roached mane who carried me through the Southern California foothills in a stiff western saddle with tapaderos and a seat belt attached to the horn. I was one of those kids who begged to go to summer camp, and cried when I had to get back on the bus to the city at the end of the week. To this day, I still know the songs by heart, how to paddle a canoe and make a mean Hobo Stew.

The memories and lessons of summer camp last a lifetime. Most resident camps offer horseback riding as an optional activity, but many specialize in riding and stable management, promising hours of "horsey" hands-on each day. Specialty camps such as these are staffed by experienced riding instructors, and usually emphasize a particular style of riding, such as western, dressage or hunt seat. For chronically horse-crazy youngsters, it's a great way to ride to their heart's content and learn about the realities of horsemanship.

But sending kids to horse camp has benefits beyond them learning how to tack up and muck out. The camping experience allows children to begin developing autonomy and a stronger sense of self-esteem. They build social skills and learn about teamwork, following directions and being responsible.

"Parents are recognizing what we in the organized camping industry have known for years," said Peg Smith, executive director of the American Camping Association (ACA). "Camp is a vital element in a child's total development, and it complements the academic skills that are learned in school with experience-based life skills."

Parents can also benefit from their children's camping experience. It gives them a chance to practice "letting go" and recharge their parental batteries. For a family considering investing in a horse, horse camp is a good intermediate step between weekly riding lessons and full-time ownership. As many horse camps include feeding, grooming and general care in their curriculum, parents can gauge the depth of their child's commitment by how enthusiastic he or she is upon return.

"Summer camp is more than a vacation for children," says Bruce Muchnick, Ed.D., a licensed psychologist who works extensively with day and resident camps. "At camp, kids learn to appreciate the outdoors, develop companionship and pick up skills (that can) remain with them into adulthood."

But choosing a camp should not be done hastily. Considerations such as cost, location, curriculum and reputation should be researched before enrolling.

How Do I Choose?

Summer camps come in two varieties: day camp and resident camp. Many public stables offer summer programs for children, some of which are sponsored through a local scouting organization or community college. They may even offer sleep-overs or field trips to shows or other barns. Day camps are generally less costly than resident camp, as they don't include room, board and transportation. Day camp is a good choice for children who aren't emotionally ready to leave home for a week or more.

Resident camps are normally located in fairly remote areas, which require that staff and campers live on site. Rustic camps provide tents or tipis with cots; more upscale sites have cabins with beds and mattresses. With more than 8,500 camps in the United States, it may seem daunting to select the one which best suits your needs. But some simple guidelines will help narrow it down.

Does your child belong to a youth group? If so, this is a logical place to start. Boy Scouts, Girl Scouts, Camp Fire Girls, YMCA and YWCA all offer regional camping opportunities. Generally these organizations offer programs geared toward a multi-activity experience which includes horseback riding. But many offer specialty programs that emphasize a particular activity. Ask if your local scout camp has a "horse unit."

Is the camp ACA accredited? The American Camping Association is a national organization of camp professionals dedicated to ensuring that camps provide a safe, high-quality level of service. To gain accreditation, camps must meet or surpass 300 national standards relating to health care, facilities, program, safety and staff training. Choosing an accredited camp means that the camp is adhering to high professional standards.

What is the ratio of campers to counselors? Personal attention is very important. The younger the children, the lower the ratio should be. For ages 6 to 8, the ACA recommends one counselor for every six campers. Ages 9 to 14 should have a counselor for every eight campers. Teenagers from 15 to 18 should have a counselor for every 10 campers.

How long is the session? Day camps are measured in hours; resident camps are measured in days. Stays of five to seven days are average, but some sessions may run two weeks or longer.

Can you visit? You may want to visit the camp prior to enrolling. Some camps offer a family visit day mid-session. It is not unreasonable to request to see the facilities.

What is the return rate? No camp is perfect for everyone, but a high return rate (campers returning year after year) indicates satisfaction with the facilities and the program. Ask for references from your area so that you can talk with other families about their impressions.

Who is the camp director? The camp director is like the school principal. He or she has a lot to do with setting the policies, philosophy and general atmosphere of the camp. Ask about the director's qualifications, including their educational background, specialized training, and how long he or she has held the current position.

What is the horse/camper ratio? Some camps assign each camper their own horse for the session. Others have to share. It is important to know how much time your child will be spending in an instructional setting each day. Three hours a day should be the minimum for a specialty camp.

What other activities are offered? In addition to riding and stable-related activities, even the most specialized camp should offer some extra-curricular activities. These may include water sports, environmental education, arts and crafts, drama or wilderness skills.

Who is the riding instructor? Counselors should be familiar with riding and instruction, but the riding instructor or head wrangler is ultimately responsible for the curriculum. Ask about his or her previous experience, accomplishments, accreditation and qualifications.

What does it cost? With average costs ranging from $10 to $50 per day for day camp and $15 to $120 per day for resident camp, there is an alternative for every budget. In some cases, parents may be able to reduce the costs by applying for partial or total subsidies in the form of "camperships." Although they are usually awarded based on need, don't assume that you make too much to qualify. According to a 1998 survey of camp directors, 65 percent of camps offered some level of financial assistance. Ask about availability, and apply early.

Beyond the cost of tuition, inquire about fees for special programs and trips, and spending money if there is a camp store.

If your child does not already own riding gear, you will need to factor in the cost of appropriate clothing, head and foot wear. Some camps provide helmets, but all should require them. Hard-soled shoes with a heel and breeches or jeans befitting the style of riding are a minimum necessity.

For those who already own a horse, you will want to inquire if it is possible for your child to bring her horse and/or tack. The facilities are an important part of this equation, as turning out a strange horse with an established herd can be disastrous. But if stalls or corrals are available, it may be feasible and even preferable to bring a horse from home.

Getting Ready

Once you have selected a camp, it is a good idea to take your child to visit the site ahead of time. Being familiar with the surroundings reduces the likelihood of homesickness.

Packing should be simple and limited to essentials. In addition to riding attire, clothes should be comfortable and durable. The staples of a camper's wardrobe include shorts, jeans, T-shirts, tennis shoes and a bathing suit. Use a permanent marker to write your child's name on every item, or iron-on name tags instead. New shoes, including riding boots, should be broken in and comfortable before leaving for camp.

Other useful items include a hat, raincoat or poncho, flashlight with extra batteries, camera with film, canteen or water bottle, prestamped and pre-addressed stationery, sunscreen, lip balm and insect repellent. Check with the camp for suggested bedding/sleeping bags, as well as any other specific gear needed for that camp. If your child takes medications, send the medicine in the original bottle, along with dosage instructions.

Camp professionals warn not to tell your child that you will "rescue" him or her from camp if they don't like it. "Why even plant that idea?" asks Meredith Sheridan, of Salem, Oregon. Sheridan is a junior high school teacher and horse trainer who has managed summer riding programs for many years. "Parents should acknowledge their child's feelings if they are uneasy, but they should also reassure them that they are going to have a great time. Offering to bail them out can become a self-fulfilling prophecy."

Summer camp can be a rewarding and educational experience for almost any child. And the lure of spending days on end riding and caring for a special horse can eclipse potential homesickness. Horse camp is a way to introduce a child to the realities of the horse they've been clamoring for, or build on an existing foundation of horsemanship skills. With professional instruction and supervised outdoor activities, it might just be the most practical way to disguise an important educational experience as summer fun.

PUMP UP YOUR HORSEMANSHIP

Jennifer Forsberg Meyer

You take lessons, attend clinics, watch videos. You read books and magazine articles. You ride regularly, keep a riding journal, and use mental imagery to reinforce what you've learned. In other words, you're doing everything possible to be the best rider you can be, right?

Nope. If you're not also pursuing physical fitness, you're shortchanging your in-saddle efforts. A fit, well-conditioned body sits a horse better, provides clearer, more consistent cues, and is less likely to be injured in the event of a fall. And weight training, one of the hottest trends in exercise today, can transform you into a strong, flexible, *more effective* rider in just minutes a week.

This article tells you what you need to know to begin pumping iron today. (It's much easier than you think, and you don't need to go to a gym.) We also outline the other basics of a well-rounded fitness program—aerobic conditioning, stretching, and proper nutrition. (See "Fitness 101".) That's because once you find out what weight training can do for you, you'll want to know how else you can improve your body. Bear in mind, though, that how you look isn't the point.

"Personal fitness is all about improving the quality of your life and the level of your performance, regardless of your sport," says Jennifer Sharpe, an American Council on Exercise-certified personal trainer and herself an avid rider. "The fact that you'll look better, too, is just a bonus."

Why Work Out?

Starting and sticking with any new program takes commitment. We're going to make it easy for you—by telling you exactly how weight training will make you a better rider. Once you check out the list of benefits, you'll be rarin' to go!

❑ **Improved posture.** Riding develops your lower body more than the upper; weight training evens things out. "Upper-body work with free weights, in particular, works wonders," notes Sharpe. "When your shoulder and back muscles are properly developed, good posture comes naturally." No more slumping, slouching, or collapsed hips—especially important in horsemanship classes.

❑ **More secure seat.** "Strong back and abdominal muscles and increased balance make it easier to sit gracefully at the lope, or to stay in position over jumps," explains Sharpe. *Bonus:* No more getting pulled forward by that overeager colt in a snaffle.

❏ **More effective cueing.** Weight training makes you more aware of your muscles and how they work. "It teaches you to isolate and use specific muscle groups," explains Sharpe. "Then, when you need to call upon those same muscles for riding, your increased 'muscle savvy' enables you to be more subtle and precise." The result: quieter, more effective legs; softer, more "feeling" hands. Plus, you'll be better able to use your seat to shorten or lengthen your horse's stride.

❏ **Enhanced relaxation.** Working out dissipates tension—in your muscles and your mind. "And when you're relaxed," notes Sharpe, "you're able to focus fully on your riding and competing, and to use your body more effectively."

❏ **Improved endurance and discipline.** Especially important for busy amateur riders, who often find themselves too tired to ride. "Sports psychology tells us that fatigue sabotages effort," says Sharpe. "Strong, fit riders don't tire as easily, and the discipline of working out makes you tougher mentally, too." You'll find yourself sticking to your riding schedule.

❏ **Injury problems.** Strong, elastic muscles, tendons, and ligaments are much less prone to injury. "Plus, not only are you less likely to fall," notes Sharpe, "but if you do fall, enhanced coordination will help you to land safely." *Bonus:* Chronic back pain, a problem for many riders, can be eliminated with judicious weight training.

❏ **Other benefits.** A confidence boost (knowing your body is strong lessens riding-related anxieties); enhanced overall health and happiness (you'll sleep better, and find yourself in a cheery mood more often); improved empathy with your horse (you'll understand why, for example, a proper warmup is so important to him, when you know firsthand why it's important to you); and you'll set a good example for lifelong health and fitness for your kids.

A Leg Up on Lifting

Sound good? Then here's how to begin. The four basic lifts will give you a taste of the benefits of pumping iron. Faithfully executed, they'll give you noticeable improvement in the saddle, especially if you supplement them with that old standby, the basic crunch (modified situp). They'll also form a core that you can build on as your enthusiasm for weight training grows.

Do this workout two to three times per week, with at least one day's rest between sessions to allow your muscles to heal and strengthen. Follow the recommendations for beginning weights (often none to start), and perform the movements very slowly. "Going slow forces your muscles to do the work," observes Sharpe. "If you lift quickly, momentum does some of the work, and your ligaments and tendons wind up catching the weight at the end of the move."

FITNESS SECRETS OF THE STARS

Here's what some top equestrians say about why they work out—and why they think you should, as well.

Steve Archer, trainer of youth world and reserve world champions in Western riding, Western pleasure, reining, and horsemanship; Richmond, Texas. Runs about 4 miles, five to six times per week.

"When I'm fit, I feel so much better—stronger, with faster reflexes. When I have a major event coming up, I bear down a little to increase my fitness level, because I really believe it gives me a competitive edge.

"Exercise also helps you to be more empathetic with your horse. After I've run 4 miles, if someone were to ask me to run another 5, I just couldn't. But sometimes we do the equivalent thing to our horses. Now, I'm more sensitive to that.

"My wife, Andrea, is even more 'hard core' about working out than I am. She walks on a treadmill twice a week, and trains with weights three times a week."

Patty Carter, winning trainer of youth and amateurs, and a judge with the American Quarter Horse Association, the National Reining Horse Association, and the National Snaffle Bit Association; Paris, Ontario. Does 200 abdominal crunches every day, and works out at a gym twice per week or so, using cardio machines for 30 minutes, and weights for 30 minutes.

"Crunches keep my back in shape for riding, and going to the gym is a mental break in addition to a fitness boost—it's nice to get away from the barn. When I'm on the road judging, I always take my workout clothes with me, because exercise helps me to stay mentally sharp.

"I encourage all my students to work out, too, and especially to tone and strengthen their upper bodies. As a judge, I know that someone who's physically fit—an athlete—always catches your attention. And I insist that my novice riders stretch before they ride. It puts them at ease in the saddle and reduces their chance of injury. This is especially important for the older ladies—my 'select group,' I call them."

Sandy Collier, 1993 World Championship Snaffle Bit Futurity winner; Buelltton, California. Runs or does aerobics with weights for about an hour four times per week.

"Working out improves your whole life: you look better, feel better, and ride better. As a bonus, increased lean muscle mass means you can eat more without gaining weight.

"I also find that being fit gives me more endurance, and makes me less apt to hurt myself. When you work around horses, there are any number of heavy, potentially dangerous jobs—such as hefting bales of hay. Being fit enables you to use your leg and abdominal muscles to lift properly and otherwise avoid injury.

"Plus, when you feel better your attitude is better, and this is a huge advantage around horses. When I'm grumpy, my horses don't want to deal with me!

"Best of all, if's never too late to start a fitness program. I got motivated on my 40th birthday, when I looked at myself in the mirror and thought, 'This isn't going to get any better.' I started working out, and by the time I was 41, I felt better than I had when I was 21."

Charlie Cole, multiple world champion Quarter Horse trainer; Chino Hills, California. Works out at a gym four to five times per week—2 days with a personal trainer, working on weight machines, and the rest on his own using cardio machines, such as a treadmill.

"Being fit is a definite energy boost. The fitter you are, the better able you are to squeeze everything you want to do into your day. I do a lot of judging, and the hotels I stay in generally have a gym area with at least a treadmill, so I can keep my program going when I travel.

"I'd say the major benefits of a fitness program to amateur riders, in addition to increased energy, are enhanced eye-hand coordination and overall flexibility. The better you are at reacting and using your body, the better rider you're going to be."

Al Dunning, world champion reiner and cutter; Scottsdale, Arizona. Does 100 crunches and various back-protecting stretches every morning; works out in his home gym 3 days per week for about 90 minutes using various cardio and weight machines, plus free weights.

"I had a chronic bad back and was going to retire. I'd been through pain clinics, including one at the Mayo Clinic, and even a double laminectomy. Nothing gave me real relief. Then, in 1991, I started working out with a personal trainer who designed a program to strengthen my back and enhance my riding. He calls his program Rider Fit (see 'Fast Track to Fitness'), and it lives up to its name! I'm stronger now than I've ever been. My back is healthy and pain-free. I'm not as likely to be hurt in a fall, and my recovery time is much faster.

"For the amateur rider, strength has so much to do with confidence and balance. If you're strong in the lower body and in your 'stabilizer muscles' [the pelvic girdle], you're going to be a more confident rider. And stretching is vital, too. For the amateur, I'd recommend stretching, abdominal crunches, and strength training for the leg and lower-back muscles—at an absolute minimum."

Suzy Jeane, president of the National Snaffle Bit Association, world champion pleasure horse trainer; Aubrey, Texas. Runs 3 miles, four to five times per week.

"Being fit makes me a stronger rider, plus it gives me the endurance I need to get through the rest of my day. Yes, it's a juggling act to find the time for running, especially this year, with my duties as NSBA president. But I've learned that when I don't do it, I tire out more quickly—mentally as well as physically.

"I usually run in the late afternoon, when the horses are being fed. One of the reasons I love running is that no one will go with me—it's my time alone. It's a mental break as well as a physical one, and I don't know what I'd do without it."

Martha Josey, longtime barrel racing legend; Kamac, Texas. Walks briskly or jogs for 15 minutes almost every day; fits stretching/flexibility exercises into her daily routine.

"I'm very big on keeping fit. To ride your best and be competitive, you have to stay in shape. I have very busy days, but my husband and I find time to walk or jog, usually in the evenings, right before bed.

"I'm also into proper nutrition—lots of vegetables and fruits—and a good vitamin/mineral supplement. As a result of all I do, I almost never feel tired anymore. And I teach all of this at my clinics, too, because training your horse is only a part of it. You have to train yourself, too."

Andrea Simons, a judge and multiple American Paint Horse Association world champion; Aubrey, Texas. Walks 2 to 3 miles on a treadmill, four to five times per week.

"Western pleasure is a highly competitive sport, and you really need the strength and endurance that exercise gives you to ride your best. I also have disk problems in my lower back, and the treadmill is what makes it possible for me to ride at all. Plus, since I quit smoking, getting my weight stabilized has been tough, and aerobic exercise really helps there, too.

"Our treadmill is in my daughter's room, and another bonus of using it is the opportunity it gives me to visit with both my daughters. We really enjoy that shared time together.

"For amateur riders, I'd say anything that can enhance the strength of their inner thigh and calf muscles is going to give them a tremendous advantage. Walking, running, squats, lunges—all are extremely helpful."

In the beginning, do only as many repetitions as you can while maintaining the proper form. If your form starts to deteriorate, stop and rest for 30 seconds, then begin again for one more set.

Before each weight-training session, warm up with a brisk 5- to 10-minute walk. After your workout, gently stretch each major muscle group, holding each stretch for 20 seconds or more. (*Note:* Before beginning any workout program, consult your physician.)

FITNESS 101

"The beauty of a fitness program," says personal trainer Jennifer Sharpe, "is that you can tailor it to suit both your lifestyle and your riding goals." Sharpe, a Northern Californian who owns three geldings and rides both English and Western, helps equestrians of all disciplines develop the strength, coordination, and endurance they need to ride their best. Here she provides some tips for designing a fitness regimen that will work best for you.

THE BASICS

Ideally, a fitness program has four components:

1. Aerobic exercise, which places a demand on your cardiorespiratory system by working the large muscles of your legs and buttocks continuously for 20 to 30 minutes or more at a time. "This is what we typically call the, 'cardio' [heart] component of a workout," says Sharpe. Examples include brisk walking, running, swimming, cycling, aerobic dancing, or using one of the popular cardio machines, such as a stationary bicycle, stair climber, treadmill, rowing or cross-country-skiing device. Ideally performed two to three times per week, cardio work strengthens your heart and lungs, burns calories, and builds your endurance.

2. Strength training, which tones and develops muscle tissue by asking it to contract against a resistance (usually a weight). Basic weight-lifting programs can be done at home or at a gym, using free weights (dumbbells and barbells) or machines. "Lifting your own body weight, as in pushups or abdominal crunches, is also strength training," notes Sharpe. Ideally performed two to three times per week, resistance training does more than increase your strength—it also improves your muscle control and coordination, fights fatigue, and revs your metabolism, causing you to burn more calories at rest.

3. Flexibility work, which stretches your various muscle groups through their full range of motion. "Stretching is best done right after your cardio or strength work, when your muscles are thoroughly warmed up," notes Sharpe. There are also entire programs, such as yoga, that focus on flexibility work. In addition to making you more graceful and limber, stretching enhances your ability to relax and helps protect your muscles, tendons, and ligaments by reducing their susceptibility to pulls, tears, and stress injuries.

4. Proper nutrition, which provides your body with just the right amount of the high-quality fuel it needs to perform at its best. Theories on how best to eat abound, but the smartest advice is the simplest: Eat anything you like in moderation, emphasizing low-fat, high-fiber complex carbohydrates, especially fruits and vegetables.

Eat only when you're truly hungry, savor your food, and stop eating when you're satisfied—not full. "It comes down to common sense," says Sharpe. "Basically, it's two cookies instead of five. If you can turn that thinking into a pattern, you've got it made."

Success Tips

The most helpful advice for starting a fitness program is, as the Nike ads say, just do it. "Start squeezing some exercise into your daily routine today, and work the details out as you go along," recommends Sharpe. Also:

- **Go easy.** "Set realistic short-term goals," she adds. "That way, you motivate yourself with success, and avoid injuries and burnout."
- **Get help.** Especially if you want to lift weights, find a professional to help you design your program, at least. Refer to "Fast Track to Fitness" for suggestions on how to find a personal trainer.
- **Make it fun.** Experiment with different forms of exercise until you find something you genuinely enjoy. If you're the sociable type, working out at a gym or attending aerobics classes may be just the ticket. If you're a loner, working out at home with cardio machines and free weights may suit you best. Walking, running, and cycling are terrific activities that can be done solo or in groups.
- **Be flexible—and creative.** Avoid rigidity. Take each day as it comes, and if you miss a workout or two, just get back on track as quickly as you can. Being flexible also means being able to substitute one form of exercise for another as needed. If rain keeps you from jogging, grab a jumprope. If travel takes you away from your regular gym, use the hotel's gym, jog in place in your room, or use the stairwell. "Exercise your creativity as well as your muscles," says Sharpe.
- **Keep going.** Make a commitment and resolve not to quit. If you must stop for awhile just start again. If you find you hate your rowing machine, trade it in for a bicycle. If time shortages overwhelm you, cut back on your total workout time—but don't give up. "If you persist," says Sharpe, "you'll get hooked on the benefits fitness provides. Then you'll be set for life."

PART III

CONSTRUCTION AND FLUID POWER

HYDRAULIC POWER RELOCATES CAPE HATTERAS LIGHTHOUSE

Richard T. Schneider

In 1870, America's tallest brick lighthouse began guiding ships at Cape Hatteras along North Carolina's Outer Banks coastline. That location later became part of the Cape Hatteras National Seashore. While modern boaters use high-tech, onboard navigation aids instead of the flashing beams from the 208-ft lighthouse—which can be seen some 20 miles away—National Park Service rangers believe that they still serve as a source of psychological comfort for sailors at sea in the area.

Over the years, Atlantic storms and surf eroded the beach in front of the lighthouse, cutting the distance between structure and sea from 1500 ft to 150 ft by 1990. The historic lighthouse was in danger of being overtaken by the waves, so the National Park Service initiated a project to move it 2900 ft to a new, safe location.

A team consisting of consultant Pete Friesen, Linden, Wash., International Chimney Corp., Williamsville, N.Y., and Expert House Movers, Virginia Beach, won the contract. They called on Jahns Structure Jacking Systems, Elburn. Ill., to devise hydraulic systems to support, lift, and move the 4800-ton lighthouse.

The crux of the move would be a *unified lift,* in which a master cylinder strokes a series of slave cylinders connected to vertical jacks positioned under the structure. Flow to each jack is identical regardless of its load, so the lighthouse would rise evenly. (Pete Friesen holds a patent on this procedure.) While the actuation involved jacks—often considered the simplest hydraulic device—the number of units, size and configuration of the load, distance to be traveled, ground conditions, and the effects of weather made the project far from simple. (For example, some of the long hose runs would expand enough under high pressure to affect fluid volume in the system after they relaxed.)

To start, engineers from International Chimney developed a shoring system at the lighthouse site. The original builders of the lighthouse had put a mat of rough-cut pine boards about 6 ft below existing ground level, beneath the water table, to support a foundation of granite slabs. The first stage of the move involved separating the structure from its foundation (with a diamond cable saw) and then removing sections of granite (using hydraulic splitters and rock drills) to make room to install shoring towers about every 8 ft. Mats of steel beams were placed on top of the pine mat to support the towers, and four 55-ton hydraulic cylinders with locking collars—from SPX Power Team, Owatonna, Minn.—on each tower were extended to take the structural load. Workers had calculated the pressure for each individual cylinder to support its intended load, and after applying the pressures, the locking collars were engaged to hold the load mechanically while pressure was released. When this phase of the project was complete, 100 cylinders supported the lighthouse while the rest of its granite base was removed.

The next step was to place the support frame (which incorporated 100 hydraulic jacks) for the acutal lift between the shoring towers and raise the lighthouse enough with the frame to lay the moving system consisting of rollers and roll beams underneath it. The unified lift came into play here. A twin-pump power unit, driven by a 180-hp John Deere engine, stroked the master cylinder to supply 4800-psi fluid to the slave cylinders. Flow from the slave cylinders was delivered uniformly to the jacks. When the slave cylinders completed their strokes, the 100 jacks were locked off. The master cylinder then retracted to replenish the system for the next incremental step of the lift.

After the moving-system structure with its jacks was in place, individual shoring posts with the SPX cylinders were re-established on steel cribbing to each side of the main beams. Next, the roll beams or tracks and roller carriages were positioned, and the building was lowered to bring the jacks into contact with the carriages. When these jacks extended to support the weight, the actual move could begin.

The circuit was reconfigured to pipe the jacks in three zones, producing a tricycle effect to minimize the potential for stressing the structure. Within each zone, jacks were interconnected as cells. The overall circuit was designed so that if a hose broke or a jack failed, enough capacity remained to provide steady support. Dozens of sensors rigged to the structure measured tilt, expansion, temperature, and vibration, and reported these conditions to the control panel.

Next, a series of pushjacks, clamped hydraulically to the main track beams, extended horizontally in unison, rolling the lighthouse along the tracks about 4 to 5 ft at a time. The track was laid on steel mats, placed on top of a layer of compacted gravel. Sections of the track were picked up and moved ahead of the structure after the lighthouse passed over them.

Twenty-three days after the move began, the lighthouse arrived at its new site—without a crack—where a concrete foundation pad awaited. The shoring towers were re-established under the structure to facilitate the removal of the moving and support frame systems, and they remained until replaced by masonry—providing additional protection against lateral movement for the next month and a half, during which they dealt successfully with the winds of Hurricane Dennis.

SERVOPNEUMATICS TAKES CONTROL

R.T. Schneider, editor

Advances in servopneumatic technology and components have opened the doors to new automation applications for these inexpensive systems.

No, that's not a typographical error in the above headline. Servopneumatics does move a head—a mechanical head, that is—across a matrix of positions to measure, cut, and imprint hoses, flexible tubing, cables, and wires. The Model 500XL fully automated cutter/printer can process as many as 40 different hose sizes in sequence. Marken Manufacturing, the machine's builder, found that servopneumatics provides a simple and inexpensive, yet highly flexible and reliable solution for the positioning tasks involved in their 500XL machines.

When it comes to designing multistep positioning systems, most engineers think of belt or ball screw drives powered by servomotors or stepper motors. Air cylinders, recognized as an ideal, inexpensive point-to-point motion-control solution, typically don't come to mind for multi-position tasks. The advent of practical servopneumatics, however, gives engineers another technology for positioning applications—particularly those that don't require the highest precision—that keeps the cost of the control system low.

Marken's 500XL machine can feed, cut to length, print part numbers, and outfeed a range of hose sizes up to 1-in. OD in a pre-programmed sequence. The machine is particularly suited for producing hose kits and wire harnesses, or for dealing with high-speed cutting and printing of different hose sizes, such as those required by the automotive industry. A key to the 500XL's success has been the simple design and operation. It is heavy-duty, easy to operate, and uses rugged components designed to last for years of trouble-free performance.

The head mechanism in the 500XL carries the drive assembly, measuring wheel, guillotine cutter, and ink-jet printhead. A Festo 2-axis servopneumatic system positions the head mechanism precisely over a matrix of hoses waiting in guideways on three levels.

The design of the servopneumatic positioning system is relatively simple. It consists of a standard Type DGPL rodless linear actuator, a position transducer, a 5/3-way proportional servopneumatic valve, and a servocontroller. The head mechanism is mechanically connected to the rodless cylinder, which can be stopped at more than 500 different positions within its stroke length. The servovalve controls the cylinder position, which in turn positions the head mechanism over the selected hose.

A transducer, mounted along the side of the actuator, provides position-feedback signals to the controller, which compares the actual position to setpoint and sends correcting signals to the servovalve as needed. The closed-loop functionality of the system yields positioning accuracies to 0.01% of stroke and repeatability of up to 0.04%—both quite acceptable for this application.

When the head is correctly positioned over the hose, a conventional air cylinder pushes down on the drive belt to pull the hose out to the length specified by the program. Another conventional cylinder with guillotine cutting blade then extends and cuts the hose, after which the hose is printed with its part number and fed out to a bin receptacle. The process then repeats for the next hose in the program sequence.

The 500XL machines are controlled by a touch-screen industrial PC (with 100-MHz Pentium processor) running on Windows 98 and a Profibus device-level network. The programs, written in Visual Basic 6.0, automatically set the length, quantity, text to be printed, and type of hose, flexible tubing, cables, or wires to be produced.

Programming the pneumatic system also is quite simple, using the dedicated Festo SPC200 servocontroller. Basic programs can be entered with a keypad. More complicated programs can be programmed with Festo's WinPISA Windows-based software on a PC—providing programming, set-up, and diagnostics capability. The controller accepts both digital and analog feedback signals and interfaces with the main machine controller via normal input/output and/or fieldbus links. On the Marken 500XL, a Profibus link signals the pneumatic valves which actuate other cylinders on the machine.

Marken Manufacturing engineers at first were skeptical about using the servopneumatic system. But after preliminary testing, it proved more versatile than traditional screw-drive servo controls. The servopneumatic drive was very smooth, cost effective, required little space, and provided the degree of precision needed for this application. Another advantage: should the machine go down, the operator needs only to shut off the air supply and reposition the linear actuator by hand, rather than back-driving a screw drive. Furthermore, the servopneumatic system uses off-the-shelf components, which can be replaced quickly if necessary.

Fast Servopneumatic System Smoothly Rejects Bad Bottles

A high-speed servopneumatic actuator system, developed by HR Textron Industrial Products, automatically rejects cans and bottles that fail inspection from beverage packaging lines. Rejection is consistent and smooth, while containers pass the actuator at rates as fast as 20 per second. The system pushes rejected containers off the packaging line without tipping or spilling them, even when they are touching and it adapts to containers of varying heights, diameters, materials, and fill weights.

The actuator system works in conjunction with X-ray and video camera inspection stations that detect container imperfections such as low liquid levels, misplaced labels, cracks, chipped mouths, or the presence of foreign objects. Designed for easy installation, the digitally controlled system consists of three HR Textron components: pneumatic servovalve with integrated electronic controller; an electronic control card; and a custom-designed low friction pneumatic cylinder with an integral linear variable-displacement transducer. The rejection system is mounted one to two feet downstream from the inspection station.

An analog signal from the inspection-station computer is received by the electronic controller. The controller interfaces with the pneumatic servovalve (mounted on the cylinder housing) and the transducer built into the cylinder. The cylinder—oriented perpendicular to the packaging line's direction of travel—simply extends to displace defective containers from the line. Load weights (excluding the container) are about one pound. Maximum velocity for the combination of piston, rod, and faceplate is 100 ips. The performance challenge, of course, is to hit the bottle squarely with just enough speed so that it slides off the line without falling over, while avoiding contact with the containers that passed inspection.

To accomplish this, the inspection-station computer is preprogrammed with motion profiles that control actuator position, velocity, acceleration, and deceleration during the rejection cycle. These profiles are determined empirically by the builder of the inspection/rejection equipment. When all containers are the same size, a single motion profile would serve the decision to reject. Displacing containers multiple distances onto multiple rejection conveyor lanes requires a separate profile for each defect. Additional profiles are needed for containers of different materials and sizes on the same line. However, only one rejection actuator system is required for each installation—regardless of line speed, container type, or number of conveyor lines.

The operator selects appropriate profiles before the inspection operation starts. Thereafter, the computer determines precisely when to send a motion-profile command. Actuator motion during the complete cycle is adjusted continuously by the electronic controller in a closed feedback loop that is independent of container speed.

HR Textron's development of the Model 27C R-DDVT11 (Rotary-Direct Drive) servovalve made the rejection actuator system possible. This design controls flow through the valve in a single stage that is independent of supply pressure. Single-stage flow eliminates springs, nozzles, filters, and the pilot valve normally found in other servovalves, significantly cutting the cost. Within the servovalve package, a limited-angle DC torque motor drives the valve spool directly through an eccentric on the motor shaft. The eccentric translates motor rotation into linear spool displacement to modulate air flow through the valve ports. An integral integrated-circuit electronic controller continuously compares an input command with spool position (determined by Hall-effect magnetic sensor). When the controller detects an error, it passes current to drive the motor and spool to the commanded position.

This servovalve's frequency response is more than 150 Hz. The high pneumatic bandwidth allows rapid adjustment of air-flow rates when the electronic controller receives a new motion profile. High bandwidth also lets the system to be tuned for maximum motion repeatability and consistency.

Other pneumatic control systems rely on conventional analog 4-way solenoid valves for motion control. These systems use mechanical stops and manual flow-control valves that must be adjusted each time the motion profile changes. This means adjusting stops and valves each time a different size container is run through an inspection head. Otherwise, containers would tend to tip and spill when rejected from the line. Further, beverage packaging lines using solenoid valves often require two or more rejection actuators to accommodate high line speeds or sorting of containers of varying size.

Motion profiles of HR Textron pneumatic servovalves are always selected through software. The number of controlled positions along the actuator stroke is virtually unlimited, allowing very precise tailoring of load position, velocity, and acceleration for a wide range of industrial applications.

Press Roll With Servopneumatic Control Provides Smooth Motion

In a sawmill, the job of a press roll is to hold a cut log (called a *cant)* against a conveyor chain to limit the cant's horizontal and vertical movement as the conveyor feeds it into a downstream saw. The design challenges in controlling a press roll include the need to carefully bring the roll into contact with the cant as it arrives under the roll's position, and to ensure that the roll follows the irregular surface of the cant as it passes. Considerable finesse is required in order to avoid banging the roll against the cant (which can damage the wood immediately, and, over time, damage the machine), while maximizing throughput of lumber into the saw.

Pacific Fluid Systems (PFS) has developed a press-roll controller that offers higher throughput and lower maintenance costs for sawmill operators.

THE 2000's FLUID POWER NEEDS
AND REQUIREMENTS

Julius Kendall

First, one must look into past years to review the products, systems and applications that have been engineered and successfully produced to meet the need of both industrial and mobile applications.

The fluid power industry has responded to these needs by developing products to satisfy the OEM's and users who specify hydraulics and pneumatics for their systems, based on inputs from their customers (users).

There has been no visible coordinated program for Fluid Power manufacturers to spend time working with the end user, to determine what these "users" would like to have in order to increase their productivity, by providing fluid power products with a higher degree of quality, reliability and simplicity. This information should be directed back to the fluid power manufacturer by the distributor, and the manufacturers' regional sales offices.

Most users who operate equipment have a degree of familiarity with fluid power components based on experience of a limited number of their employees, including engineers, maintenance personnel and machine operators. This experience and knowledge is based on their personal "hands on" working with components to solve functional problems, failures, repairs and/or modifications.

The OEM manufacturers respond to the users needs through the contacts made by sales and engineering personnel with the customer's management, production personnel and facility engineers. These people all influence management's decision as whose equipment will be purchased, and which components they would like to have.

There are thousands of components in use today on a broad spectrum of equipment in every conceivable industry and application. These products are the frontline "salesman." Their performance and reliability help sell future products from the same manufacturer.

The sale of spare parts to a large measure helps finance the development of new and unique products needed by industry. Such sales represent a sizeable percentage of total sales, at higher gross profit margins.

From *Fluid Power Journal, Certification Directory, 2000* by Julius Kendall, PE, CFPE, Vice President—The Entwistle Company. Copyright © 2000 by The Fluid Power Society. Reprinted by permission.

What are these products? Many experienced and well known persons have expressed their views as to what will be needed to satisfy their needs, be competitive, become more global oriented and aware as to the ready response that customers demand.

The Fluid Power Industry is only one persuasion looking to the future, and what will be needed to compete. Many different subjects have been expressed, and in most cases, good inputs are available worldwide. It is a global business.

Industries are examining new materials, new fluids, better ways for products to communicate with each other, alliances with electronics, electrical, programmable controls, artificial intelligence, feedback methods, self testing, etc.

The end user of tomorrow will be more dependent on visual indications as to how well a system works, and if there is a problem, they expect the problem to identify itself, and indicate the corrective action to be taken. End users may not be as skilled in "trouble shooting" a system or product, and therefore expect the manufacturer to provide the solution.

In talking with several chief engineers of companies who are OEM to industry, they were asked what do you need for the future to be competitive?

Generally they indicated:

1. Better quality
2. More reliability on performance
3. Noise reduction
4. Simpler products
5. Competitive pricing
6. Better technical information about the product, its functions, installation and expected performance

The following are various comments and suggestions that were offered by companies and personnel contacted, as well as personal observations from contacts with DOD agencies, etc.

a. Higher operating pressures, up to 8000 psi
b. Higher operating temperatures
c. Better and less costly hydraulic fluids
d. Eliminate external and internal leakage conditions
e. Better and simpler type connectors—fittings, hose
f. Investigate the use of better, non-corrosive materials such as ceramics, glass, etc., that are not subject to corrosion or need lubrication
g. More consideration to use cartridge type valve and manifold assemblies
h. Lower energy consumption
i. Proportionally controlled valves, pumps, etc. user friendly, all integrated

j. Increased operating time, without loss of efficiency
Basically, most OEM's break their requirements for systems into three categories:

a. Power—pumps, etc.
b. Control—relief valves, pressure valves, etc.
c. Motion—motors, cylinders, etc.

To establish needs on a global basis for the future, one must also address the source for engineering talent, technicians and skilled mechanical personnel.

The Fluid Power industry has made an effort by helping to establish educational programs worldwide by working with colleges and technical schools, to include fluid power courses in their curriculum. This program is not as successful as the industry expected. The fluid power industry must reach lower into high schools, and educate both teachers and students about fluid power to create a demand for higher education when these students apply to colleges.

Universities are also businesses, and they offer courses that are in demand and popular with incoming students. If no interest, then the University does little to offer any course that will not be profitable. This brings up many questions:

1. Why are university students not interested in fluid power? Should fluid power be linked to electronics, computers, etc. to show the link?
2. Why is the fluid power industry so misunderstood by the public? Do we lack visibility?
3. Why hasn't Wall Street been more educated about the public companies in Fluid Power, their contributions, their applications, etc. This could afford more investments and attract more people to our industry.

Another consideration in looking ahead is what can we expect from our labor force. The direct labor work week is shrinking from the conventional 40 hour week to 35 hours. The cost to support labor has not changed. Direct labor wants to maintain their standard of living, and therefore will look for second jobs, working 25 to 30 additional hours a week.

Will they have the same loyalties to their primary job? Will they be just as efficient? What will industry do to insure that products maintain a high degree of quality under such conditions, and how will performance be measured?

Today's product catalog contains many components that have been offered five years ago. These products for the most part are still very much in demand, and will still be in demand five years from now.

It is the consensus of opinion of many contacts including many well known consultants, that in present catalogs, 92% of the products offered today will still be in demand in 1999 and beyond.

At the same time, there is and will continue to be systems that demand the use of electronically controlled valves and pumps, using proportional or servo controls. The fluid power industry will rise to the challenge and either develop their own electronic controls, or acquire such controls from the electronic industry.

The configuration of such electro-hydraulic devices will depend on the system concept that the OEM requires for the total product arrangement, and the ability and means to interface, in a simple and friendly manner.

Evaluating current sales will indicate those catalog products that major OEMs are still specifying for their systems, and provide the desired performance for their equipment.

Based on a continued use of 92% of today's available product, the requirements of more intelligent electronically linked components are for systems that the OEM is attempting to introduce into the marketplace, that will compete head on with both domestic and foreign manufacturers.

Electrically operated and electronically controlled systems are more evident in many areas, simply because they can be applied and installed at lower costs, and can readily be linked with the customer's computer network, or the equipment can easily be set up with a local key pad, and the operator trained.

The Fluid Power manufacturer has permitted such products to enter the traditional market by responding to the OEMs technical requirements by offering products that will perform more efficiently, quieter, cost effective and technically equal. Electronic drive systems for the most part do not perform as well in higher H.P. hydraulic drive systems, nor have the versatility.

The selection of electronic drives by OEM engineers seem to be directed by electronic rather than hydraulic engineers. Are the right people being contacted?

The fluid power industry accepts the advantages of using and linking electronics in its many forms to hydraulic products. It provides the user with means to communicate control and motion when and how desired. These are attachments that are integrated into fluid power products, where the product is being redesigned or modified to incorporate these electronic features.

What the OEMs need are products that are more energy efficient. Utilizing the cartridge concept in manifold blocks, with proportional type controls mounted directly to the motion device, eliminating interconnecting piping, tubing and/or hoses, thus reducing potential leaks, improving response time and incorporate such electronics that are needed for the system.

The requirements for future fluid power products vary with the market. The mobile market has constantly demonstrated a need for remote-controls, and industry has provided products to meet such needs. The industrial market has been much slower to accept the need for such products in the past, but are now being forced to consider remote controls that can be system integrated.

Cartridge valve products have been available for many years. These products are now receiving more attention, both in Europe and in the States. The ability to provide integrated manifold designs incorporating a variety of functions including flow, pressure and direction in a single manifold and attaching it to the motion device offers many advantages.

Summary

Marketing has determined that there are products that the OEM needs for their system, these include more efficient piston pumps incorporating electronic controls based on developments to date, electronically operated controls for flow, pressure and motion, modular cartridge packaging integrated to the motion device.

Products offered must be quieter, more energy efficient, user friendly, self testing, be more dirt tolerant and cost effective.

Many consultants and those coordinating with NFPA have expressed their views regarding products for the 1990's with more consideration for modular cartridge packaging and direct attachment to the motion device, for quieter systems, more energy efficient and collaboration with electronics.

It is the collected opinion that the majority of products presently cataloged will still be in strong demand for at least the next five years.

Management should make long term commitments to develop new products that will be specified by the OEM and the government, such as:

1. better materials like ceramics, glass composites
2. better and less costly hydraulic fluids
3. dirt tolerant components
4. more friendly electronic controls
5. self-testing and identification of problems, and
6. longer operating life.

More efforts should be made to work closely with major OEMs by either offering to integrate a new product into their system to prove its value, or purchase a machine and make the installation and prove the value by actual demonstration and field testing by a selected user.

Consider investigating the use of ceramics and/or glass in pumps and valves could help reduce cost, increase efficiency, reduce lubrication problems, increase life and reduce weight.

The user is a most important link to future product development. They are the source for continued spares and spare parts business. They should be encouraged to purchase ONLY genuine original manufacturer's specifications, and offer such repaired components to the user with a limited warrantee, and at a fixed exchange price. Promoting at all times "Genuine Replacement Products."

The establishment of rigid NFPA and ISO specifications has definite restrictions of product development. Some users demand interchangeability between manufacturers of the same product. This results in price becoming the major factor rather than quality, reliability, reputation, etc.

Developing the cartridge type manifold concept facilitates providing products that can be engineered to better satisfy the application, at a lower cost, user friendly and with a better link to the after market. It also offers the OEM a more proprietary product for his protection.

What also needs to be considered is more friendly and easier to use catalogs. The design engineer may have more familiarity with the catalog, since it is used with more frequency. Yet, due to the amount of technical data that is provided, it is difficult, in many instances, to find the essential technical details, resulting in calls to the factory or the local regional office or distributor.

The user, the mechanic, plant or facilities engineer, who does not use the catalog with any frequency, finds the technical information difficult to understand, and in some instances, the exact nomenclature is not shown in the catalog, resulting in further frustration.

The technical needs of the OEM, user, etc. should be reviewed and possibly there may be a need for two types of catalogs. One specifically for design and engineering personnel, and one for the end user, who needs to identify the product based on the nameplate data, and be able to understand its function to purchase replacements or parts.

With the user of CD disks, such as some manufacturers are now offering, complete with symbols that could be used with Cad Cam as well is a step in the right direction. But, what about the user's mechanics, who want a simplified technical catalog that is prepared in PLAIN language with "lots" of sketches, photos, etc. They do not have the time to read complex technical manuals.

This report was prepared in response to a need observed by major fluid power manufacturers as part of a program for "What will industry require of the Fluid Power Industry into the year 2000?"

LIVING WITH PORTLAND'S URBAN GROWTH BOUNDARY

Jon Vara and Martin Holladay

When Portland, Ore., put limits on urban growth in 1979, the measure enjoyed broad public support. The Urban Service Growth Boundary, or UGB, was designed to preserve farmland and prevent urban sprawl by imposing a strict outer limit on residential development. But this year, as the UGB approaches its 21st birthday, area builders are severely pinched by a shortage of land within the boundary, and many are calling the policy deeply flawed.

At the heart of the controversy is Metro, the regional agency responsible for land-use planning in the area enclosed by the UGB, which encompasses parts of three counties, the city of Portland, and 23 of its suburbs. As the region's population has grown, Metro has responded to the demand for new housing by mandating smaller lot sizes and promoting row houses and other high-density housing, such as combined residential and office space. That has kept housing available, but at a cost: Portland now ranks as one of the nation's least affordable cities.

Jeff Fish is a Portland contractor who builds 20 to 25 houses per year, mostly as infill. "We bought into the boundary idea because we didn't want to see all the land go for strip development," Fish says. "Now we're all clamoring after the few lots that are left." Today's builders, he says, are paying nearly $50,000 for building lots that went for $7,500 just seven or eight years ago. "If a lot goes on the market at eight o'clock and you don't get there until noon, you find that there are already four or five offers on it."

Planning advocates, on the other hand, contend that the scarcity of land means that the system is working as it was meant to. Conditions within the UGB, they say, are no different from those that confront builders in older, heavily developed metropolitan areas elsewhere in the country.

Large builders, however, who need economies of scale to operate at a profit, have been hurt badly by the growth restrictions. "The average subdivision inside the UGB is around 19 units," says Kelly Ross, of the Home Builders Association of Metropolitan Portland. "In most parts of the country, it's more like a hundred units. Building is becoming a sort of boutique industry."

Many smaller builders are also looking beyond the Portland metropolitan area, where land is easier to come by. "There's a lot of building going on around McMinnville, down in Yamhill County," says Kelly Ross. The state capital of Salem,

---

Census Bureau Releases Top 10 List

Recently released housing growth statistics from the U.S. Census Bureau paint a revealing picture of building activity over the past decade. All of the top ten states for building fell into two distinct groups—the first in a band stretching south and eastward from the Pacific Northwest, and the second in a more compact cluster in the Southeast.

The Bureau's numbers on housing units, which covered the period from 1990 to 1998, found that Nevada led overall, with a whopping 48% increase. It was followed by six other western states: Utah, Idaho, Arizona, New Mexico, Washington, and Oregon.

Much of the region's growth is attributed to a steady influx of retirees into the desert states. Nevada also posted the largest percentage increase in households aged 65 and over, with Arizona and Utah registering third and fifth in that category.

---

40 miles south of Portland, has also been a focus of building activity. Ironically, commuter traffic between those fast-growing areas and Portland itself has become a major source of traffic congestion—something that the growth boundary was intended to prevent.

Others are going even further afield. Jeff Fish has been selling some property in Portland and buying land in Las Vegas. "I think I'll be building down there a year from now," he says. "A lot of us are really struggling. I know two guys who are building in Phoenix, and four or five of my competitors are already in Vegas. It's not a matter of wanting to go, but of being forced out."

Those home builders who remain are often left grappling with policies that seem designed to frustrate them. The city's emphasis on townhouse developments can spell disaster for a builder who misjudges the market—because, builders say, it's impossible to force customers to buy products they don't want. "You can sell $120,000 townhouses in the suburbs," says builder Ron Nardozza, "but nobody wants an expensive townhouse unless it's right in town. I can show you some very nice $225,000 townhouses that are just sitting empty." Bob McNamara, a land-use planner with the National Association of Home Builders, agrees. "No one really wants to live at eight units per acre. It's something people put up with when they have no choice."

Under state law, Metro is required to adjust the UGB as required to maintain a 20-year supply of building land. In 1998, Metro sought to push the original line out to encompass an additional 3,500 acres. But in the pressure-cooker atmosphere of Oregon land-use planning, it soon became clear that altering the boundary would be much more complicated than simply changing a line on a map.

Conservationists and farm groups oppose the boundary expansion, as do the wealthy residents of rural "hobby farms" outside the existing line. A coalition of expansion opponents has gone to court in an attempt to block it. That case may take several years to resolve, and has prompted the Home Builders Association of Metropolitan Portland to launch its own legal challenge, aimed at forcing the state to live up to its obligation to make the land available. Meanwhile, disgruntled citizens—angry at Metro for imposing row houses on what were traditionally single-family neighborhoods—are threatening a ballot initiative to strip the agency of its power to establish minimum densities.

But while the growth boundary has been a major headache for builders of new homes, it has actually improved business for remodelers, at least in the short term. "Business has been good lately," says Mike Kelly, of the Neil Kelly Co., a family-owned design-build/ remodeling business. Because the UGB tends to drive up property values and restricts new houses to small lots, he explains, many homeowners would rather invest in a major remodeling project than move into a new home. "You have people paying $200,000 for a house, then spending another half-million to remodel it," Kelly says. "Wholehouse remodeling probably makes up 15 to 20 percent of our business today. Ten years ago, it was more like 5 percent."

Still, he cautions, even those who benefit from the current situation aren't necessarily happy with it. "We're all in this together," he says. "A lot of remodelers build a new house now and then, and most builders do some remodeling. Without a viable homebuilding industry, remodeling has no future."

ONE-PIECE STEEL FRAMING SYSTEM MAY
REVOLUTIONIZE RESIDENTIAL CONSTRUCTION

A unique light gage steel framing system has debuted in the U.S. with the promise of simplifying residential steel construction in a substantial way.

Called the Scottsdale System by its originators in New Zealand, the framing elements are made from 24-gage, high tensile steel combined with a software package that is powerful, accurate, yet simple and quick to operate. More importantly, the software virtually eliminates all manual dimensioning once the design is completed in Chief Architect (by ART, Inc.), a U.S. architectural software package.

What makes the system stand out is its simplicity. In less than a few hours a design can be converted in the Scottsdale CAD program. The completed file is then exported and every element of the design is automatically and accurately produced by a CNC-controlled rollformer.

The rollformer is compact, accurate and produces elements quickly. More importantly, every element is cut exactly to length, rivet holes are punched and all the necessary functions required to fabricate the elements are automatically performed by the rollformer. This is accomplished with a high degree of accuracy—better the 1/64" on an 8' stud. The rollformer even dimples every rivet hole at every intersecting connection, resulting in very strong joints and a smooth wall surface.

The track, studs and headers are all made from the same C-section. The roof system is made up of ceiling and roof panels, again constructed using the same C-section as the wall studs. This eliminates the use of trusses and costly connections associated with them. The fabrication of the frames is done at the end of the rollformer machine.

Newly-formed SteelMicroSystems, Modesto, CA, headed by Gary Johnson, is marketing the Scottsdale System here in the U.S.

"The key to it all," says Johnson, "is software which is easy to operate. The CAD program uses Chief Architect for the initial input, including window and door openings, which is much easier than other commonly used software. The next element, called G-CAD, locates and lays out all the connections. The manufacturing component is an uncoiler and a rollforming machine. The final component is a laptop computer to run the system. The total area required for setup is six by eighteen feet. It's incredible how simple and compact the system is, but it works better than anything out there," Johnson adds.

"What takes a normal CAD system a week and half to perform, Chief Architect and G-CAD do in four to five hours," Johnson notes. "Instead of a $1,000 labor CAD cost you have from $100–$300 and you are really competitive with steel versus wood," he adds.

As a result of using high tensile steel in conjuction with the C-section design, the metal thickness is only 0.022 mil thickness (about 24 ga.), thereby saving at least a ton of steel per home. All the walls, including interior walls are load bearing. This allows all loads to be distributed throughout the house.

Currently launching the Scottsdale system here is American Steel Frame, Stockton, CA. American Steel expects to have plan approval and start building its first unit in early May.

Meanwhile, back in Australia, Melbourne-based construction company Fletcher Challenge is turning out 40 homes a month using the system. Johnson became a believer when he visited the Fletcher Challenge projects underway in Sydney.

"I saw a number of homes and apartments under construction and then went to their plant to watch the rollformers in action. The more I saw the more excited I got about the Scottsdale System. It has been 30 years since I have seen homes this easy and fast to build," he adds.

Johnson next went to Sydney to see what was going on in light gage steel since his last visit there nine years ago. A light steel engineer offered to show him some projects under construction and some manufacturing plants.

At the end of the day Johnson visited a plant about an hour out of Sydney. He recalls, "There the plant owner told us that there was a steel house being built a short distance away. When we went to it, it turned out to be a Scottsdale System house and the contractor was there.

The contractor told Johnson that he first started using steel to build houses eight years ago, but switched to timber (that's dimension lumber in Australian) five years ago because steel was "just too complicated."

Johnson says, "This was his third Scottsdale house and he said, 'I would have stuck to steel and not gone back to timber if the Scottsdale System was around five years ago!' On the drive back to Sydney, the engineer who was hosting me called someone he was working for and told him they should take a look at the Scottsdale System" The rest is history.

Gary Johnson was founder and president of Innovative Steel Systems, Lathrop, CA, the largest in-plant steel framing and erection firm in the state.

Masonry Contractors Beware

Walter Laska

The Masonry Standards Joint Committee (MSJC) last year published the 1999 edition of the ACI 530.1/ASCE 6/TMS 602 *Specification for Masonry Structures* (MSJC Specification). This document is incorporated by reference into the ACI 530/ASCE 6/TMS 402 *Building Code Requirements for Masonry Structures* (MSJC Code) and, therefore, into the model building codes *(Uniform Building Code, National Building Code,* and *Standard Building Code)* and the building codes of most states. One new provision that will greatly affect masonry contractors is covered in the Project Conditions section under Hot Weather Construction (Article 1.8 D). This new section requires that when the mean daily temperature exceeds 100° F or exceeds 90° F with a wind velocity greater than 8 mph, new masonry construction must be fog-sprayed until damp—and this must be repeated at least three times, a day until the wall is 3 days old.

Fog-Spraying Concrete Masonry Can Be Beneficial

The National Concrete Masonry Association (NCMA) has conducted research that addresses fog-spraying of concrete masonry construction. A substantial amount of test data has been collected and incorporated into a report titled "Research Evaluation of the Flexural Tensile Strength of Concrete Masonry." One of the conclusions of this research is that fog-spraying concrete masonry construction can increase bond strength in very hot and arid conditions. In addition, fog-spraying concrete masonry under these conditions can reduce shrinkage cracks in mortar.

No Data Supports Spraying of Clay Masonry

Unlike the extensive research provided by the NCMA research report, no such supportive data currently exists for spraying clay masonry. The only extensive research addressing wet curing of clay masonry was performed nearly 70 years ago for the Clay Products Institute of California. The research concluded that wet-curing merely saturates masonry, decreasing adhesion between mortar and brick, and that it appears there is little to gain by wet-curing. Currently the Brick Industry Association

Tech Note 8 states that "wet-curing of masonry generally produces higher bond strength than dry curing"; however, there is no test data to back this up.

ASTM C 270, "Standard Specification for Mortar for Unit Masonry," in Appendix X1.9.5.4 states that "with very rapid drying under hot, dry and windy conditions, very light wetting of the in-place masonry, such as fog-spraying, can improve its quality." However, no distinction is made between clay and concrete masonry. The appendix further states that "curing of mortar by the addition of considerable water to the masonry assemblage, however, could prove to be more detrimental than curing of mortar by retention of water in the system from its construction. The addition of excess moisture might saturate the masonry, creating movements which decrease the adhesion between the mortar and masonry unit."

Another potential problem with fog-spraying is that introducing excessive moisture may contribute to efflorescence.

Other MSJC Oversights

In addition to the limited test data, the MSJC overlooked other critical factors pertaining to adding moisture to a clay masonry system:

- ❑ Certain vital physical properties of the clay masonry were ignored, such as unit absorption or initial rate of absorption (IRA). Clay masonry units with very low absorption rates (IRA less than 5 grams/30 minutes/30 square inches) will extract very little moisture from the mortar. Introducing additional moisture into a very low absorption unit could further decrease bond.
- ❑ The humidity that the assemblage is exposed to has not been taken into account. Although the mean daily temperature might exceed 90° F, the humidity might also exceed 90%. Humidity this high will reduce moisture loss from the masonry system. This type of weather condition is common in the Midwest and the southeast United States in midsummer.
- ❑ Possibly the most significant problem with the MSJC Specification's requirements for hot-weather construction is the failure to recognize that these are minimum requirements for masonry construction. Although most regions of the United States rarely encounter mean daily temperatures of 90° F with a wind velocity greater than 8 mph, this does not prohibit the designer from lowering this threshold and specifying fog-spraying in milder weather conditions. Moreover, the designer could also require masonry systems to be fog-sprayed at a greater rate than the MSJC Specification requires. Given the traditional over-conservative approach of designers, fog-spraying could become standard practice on masonry construction in the summer.

Ramifications

Two major problems could occur as a result of requiring clay masonry assemblages to be fog-sprayed. First, at any temperature, introducing excessive moisture into low-absorption clay masonry units and moderate-to-high water-retentive humidity can cause lack of bond and separation cracks due to planes of water that develop between the mortar and the unit. These cracks are often misinterpreted by the designer as workmanship flaws.

Second, most masonry contractors are not familiar with fog-spraying or equipped to fog-spray masonry. So fog-spraying can easily be overlooked or improperly applied by the contractor, especially if the designer tightens the requirements. Also, fog-spraying is not well defined. No criteria have been established for how wet the units should be.

As a consequence, the contractor (even if not at fault) could be required to tear down the masonry and reconstruct it. However, the Code is the Code, even if it contains major oversights for contractors. So, "gentlemen, start your foggers."

SQUARE DOWELS CONTROL SLAB CURLING

Ernest K. Schrader

Curling at floor joints is a common problem. However, field observations and experimental studies have shown that substantial reductions in curling can be achieved by using square steel dowels.

Origins of the Square Dowel

In 1987, I proposed replacing traditional round steel dowel bars with square dowels to provide better load transfer at joints (see reference). The square shape also allows the dowels to be fitted with a clip-on device, that puts the sides of the dowel in contact with a compressible material. This permits lateral movement due to concrete shrinkage or contraction and reduces the cracking caused by restraint of lateral movement, which is likely to occur with conventional round bars. The clip is made of a hard ABS plastic, with the compressible material affixed only to the clip's sides. Vertical load-transfer forces are transferred to the square dowel by the ABS plastic, which has a compressibility similar to that of concrete.

Because the side material is compressible, the square bars can be misaligned in the horizontal plane without inducing tensile stresses in the slab when the joint opens. Instead, the side material compresses. Benefits include less corner cracking and less restraint stress that can cause cracking parallel to the joints.

Engineers who used the square dowels soon noted another benefit—reduced joint curling. The reduction was greater than could reasonably be explained by the greater stiffness of the square bars, so I devised a series of full-scale tests to validate or disprove the field observations.

The Test Program

I performed the tests on 6-inch-thick concrete slabs with a 28-day compressive strength of 4900 psi and 28-day shrinkage of about 0.06%. Slabs were placed on 3-inch-thick boards of expanded polystyrene to give minimal base support. I tested three different joint configurations:

❑ Undoweled joints
❑ Joints with 3/4-inch-diameter smooth, greased steel dowels spaced at 12 inches
❑ Joints with 3/4-inch-square ungreased steel bars equipped with the dowel clip and also spaced at 12 inches

Both round and square dowels were embedded 9 inches into the slabs on each side of the joint. I first applied increasingly higher vertical loads on one side of the closed joint through a 36-square-inch pad to simulate the contact area of a rubber tire. As expected, the undoweled joint provided hardly any load transfer, causing faulting (vertical displacement) at the joint. Both the round and square dowels initially controlled the vertical displacement.

Next, I used a hydraulic jack to pull the slabs apart and create a 0.1-inch joint opening before again applying increasing vertical loads on the slab side opposite that of the original loading. Under the open-joint condition, especially at higher loads, the round dowels bent, allowing the slabs to curl, or rotate significantly around the joint axis. At the same higher loads, the square dowels with dowel clips provided much more resistance to rotation (curl), allowing adjacent slab surfaces to remain flatter. The table shows slab rotation in degrees for the three test conditions and for open and closed joints. Readings from a series of dial gauges mounted along the slab surface measured the rotation. The table also includes an improvement factor. A factor of 10, for instance, means there was 10 times less curl or relative rotation of adjacent slabs.

The Advantages of Curl Control

This series of tests validated field observations of reduced curl when square dowels are used with dowel clips to provide load transfer. The benefits from reduced curl include:

❑ Flatter floors or pavements
❑ Less likelihood of cracking caused by loss of subbase support
❑ Increased resistance to fatigue failures caused by repeated loads

Some specifiers of square dowels now cite reduced curling as their principal reason for using these devices.

Other Advantages of Square Dowels

For slabs on a sand base, additional tests showed a 35% increase in load-transfer strength at joints with square bars instead of round bars. When the clip was added to the square dowels, it produced an additional 52% increase in strength. This increased

| Load (lb) | Joint description | | Relative slab rotation (degrees) | Improvement factor** | | |
	Type of dowel	Opening (in.)		Round vs. no bars	Square vs. no bars	Square vs. round bars
			DOWEL BENDING SUMMARY*			
			[Relative curl or rotation of slabs on each side of the joint]			
0	None	0	0	–	–	–
4,122			5	–	–	–
5,153			10	–	–	–
6.183			12	–	–	–
0	Smooth round	0	0	–	–	–
4.122			1	5.0	–	–
5,153			5	2.0	–	–
8,244			7	–	–	–
0	Square with clip	0	0	–	–	–
4,122			0.5	–	10.0	2.0
5,153			0.5	–	20.0	10.0
8,244			0.5	–	24.0	14.0
0	None	0.1	0	–	–	–
2,061			6	–	–	–
5,153			11	–	–	–
0	Smooth round	0.1	0	–	–	–
4,122			14	–	–	–
5,153			18	–	0.6	–
8,244			19	–	–	–
0	Square with clip	0.1	0	–	–	–
4,122			2	–	–	7.0
5,153			4	–	2.8	4.5
8,244			8	–	–	2.4

*3/4-inch round and square smooth bars, embedded 9 inches into each slab (6-inch-thick slab; 4000-psi concrete)

**Improvement factor = rotation for condition 2 ÷ rotation for condition 1 load-transfer capability occurred when the clip was made with a specially engineered plastic material but not with a softer plastic.

I also measured resistance to lateral, or sideways, movement. To minimize cracking, it's desirable to allow adjacent slabs to expand and contract at different times and rates, especially at corners. This isn't possible with traditional round dowel bars. When round bars are used, lateral movement or shrinkage of the slab on one side of the joint induces stresses or similar movement in the slab on the other side of the joint. However, using square bars with a compressible material on the sides eliminates this movement. Tests showed total resistance to side movement with traditional round bars and virtually no resistance to side movement with the square bars and clips.

❏ References

Ernest K. Schrader, "A Proposed Solution for Cracking Caused by Dowels," *Concrete Construction,* December 1987, pp. 1051–1053.

PART IV

ENVIRONMENTAL RESOURCES

Factors Influencing the Profitability of Precision Farming Systems

M. T. Batte

The emerging site-specific management (SSM) technologies will allow farmers to manage each parcel of land in a manner consistent with its unique endowments. The consequence of reducing the scale of land area that is managed uniquely is to substantially reduce the number of cropped acres for which inputs are either over- or under-applied. The potential impact of SSM on farm receipts, variable input costs, and fixed investment costs are explored. The likely impact of farm size on profitability of SSM adoption also is explored. A hypothetical case is explored to demonstrate the economic value of various components of SSM. Finally, the potential environmental consequences of SSM for society are explored.

Site-specific farming is an emerging technology with substantial promise to aid both farmers and society. These methods, also variously referred to as *Precision Farming, Variable Rate Application Farming, and Prescription Farming,* allow application of inputs to a specific cropland area based on soil type, fertility levels, and other endowments of that site. The site-specific management (SSM) concept is based on the ability to repeatedly locate a position within a field. Site-specific farming incorporates four technologies: Remote sensing, geographic information systems (GIS), global positioning systems (GPS), and process control. Remote sensing data have been used for several years to distinguish crop species and locate stress conditions in the field. Other remote sensing applications of SSM include yield monitors, moisture sensors, and soil nutrient sensors. GPS is a navigation system based on a network of earth-orbiting satellites that lets users record near-instantaneous positional information (latitude, longitude, and elevation) with accuracy ranging from 100 m to 0.01 m (Lang 1992). GPS allows the manager to reliably identify field locations so that inputs can be applied to individual field segments based on performance criteria and previous input applications. GIS technology allows the manager to store field input and output data as separate map layers in a digital map and to retrieve and utilize these data for future input allocation decisions. Process control technologies allow information drawn from the GIS to control processes such as fertilizer application, seeding rates,

From *Journal of Soil and Water Conservation,* Vol. 55, No. 1, First Quarter 2000 by M. T. Batte.

and herbicide selection and application rate, thus providing for variable rate application technologies (VRT).

The National Research Council Committee on Assessing Crop Yield: Site-Specific Farming, Information Systems, and Research Opportunities, defined site-specific management as "a management strategy that uses information technologies to bring data from multiple sources to bear on decisions associated with crop production" (p. 2). They correctly recognized that SSM is not a single technology, but rather a suite of technologies that allow 1) capture of data at an appropriate scale and time, 2) interpretation and analysis of that data to support a range of management decisions, and 3) implementation of a management response at an appropriate scale and time.

In the remainder of this paper, recent research into farm-level economics and environmental impacts of SSM will be reviewed, and those factors that are likely to impact the profitability of the farm firm with adoption of SSM will be identified. The likely impact of farm size on the cost of owning and operating the SSM system and the potential impacts of widespread adoption of SSM on the size structure of agriculture will be explored. A hypothetical example to explore the economic value that might be derived from grid soil sampling and variable rate application of fertilizer will also be presented and potential environmental benefits will be suggested.

Previous Research

Swinton and Lowenberg-DeBoer compared profitability estimates for variable rate application of fertilizers in nine studies, which represent profit comparisons for 54 sites. Information (soil sampling) costs were added for those studies that did not recognize this charge. They reported that 57 percent of the sites studied produced greater profits for the SSM regime than for a uniform rate technology (URT).

Babcock and Pautsch evaluated the profitability of VRT nitrogen fertilization relative to a uniform rate application in 12 randomly selected Iowa counties. They concluded that the economic and environmental impacts of moving from URT to VRT depended heavily on the inherent yield variability in fields. They found modest increases in returns above fertilizer costs. The majority of this benefit was for a reduction in fertilizer costs rather than an increase in yields. They suggested that reduced nitrogen usage will correspond to a reduction in nitrate leaching and associated external costs.

Watkins, et al. performed a similar analysis of farm and environmental economics of VRT application of nitrogen fertilizers in seed potato production in Idaho. VRT fertilizer applications resulted in decreased profits relative to conventional fertilizer application strategies. They estimated that nitrate losses from the field were virtually the same for VRT and URT application methods.

Oriade, et al. studied weed control with post-emergence herbicides. They evaluated two application techniques: uniform rate application of the herbicides over the full field and uniform rate application of herbicides for selected portions of the field (spot spraying). They also considered alternative levels of weed populations and weed *patchiness*. Weed patchiness was the most important factor influencing the profitability of SSM of herbicides. They also considered the environmental consequences of the two management methods. Environmental benefits of SSM grew with greater weed populations and increased patchiness of the weeds.

Farm-Level Economic Impacts of Site-Specific Farming

The adoption of site-specific farming practices will likely result in significant economic consequences at the farm level. However, because these technologies are just now emerging, we do not have solid estimates of resulting changes in costs and returns. Furthermore, farmers and industry personnel are still learning how to implement these systems. Thus, we expect the costs and returns to change over time as the technology and our understanding of how to use it changes.

A simple economic model can be used to demonstrate the potential for changes in both revenues and input costs. Figure 1 represents the marginal productivity of alternative levels of an input in production of some crops. The marginal value productivity (MVP) of the variable input represents the additional value (product of added yield and price) resulting from the last unit of input. The MVP curve in figure 1 is downward sloped as suggested by the law of diminishing returns. Marginal input cost (MIC) represents the cost of an additional unit of the variable input. The economic efficient (profit maximizing) level of input usage (I*) is identified by the intersection of MVP and MIC. This is the point at which the value produced by the last unit of input is exactly equal to the cost of the last unit of input.

Figure 2 extends this model to consider two soils with differing productivity with respect to the variable input. Soil A has greater productivity than soil B—MVP_A is greater than MVP_B at all levels of input usage. I_A* and I_B* represent the optimal allocation of input on soils A and B, respectively. If a single soil test is made and a single rate of input is applied, an uneconomic allocation will be made for a portion of the land base. If the soil test is drawn from soil A, I_A* will be applied to both soils. This represents an over-application of the input on soil B. In the event that the soil sample is drawn from soil B, inputs are applied to all soils at the rate I_B*, resulting in an under-application of the input on soil A. In the event that soil cores are drawn from both types of soil and mixed prior to determining the input rate recommendation, the resulting rate will be between I_A* and I_B*, and will represent the optimal input usage on neither soil.

SSM recognizes the different input requirements of the two soils. Provided that the two soils are identified correctly and soil samples are drawn from each, an optimal input allocation would be made for each soil. Compared to applying I_A* to both soil types, SSM results in a decrease of input cost (rates are lowered to I_B* on soil B), a

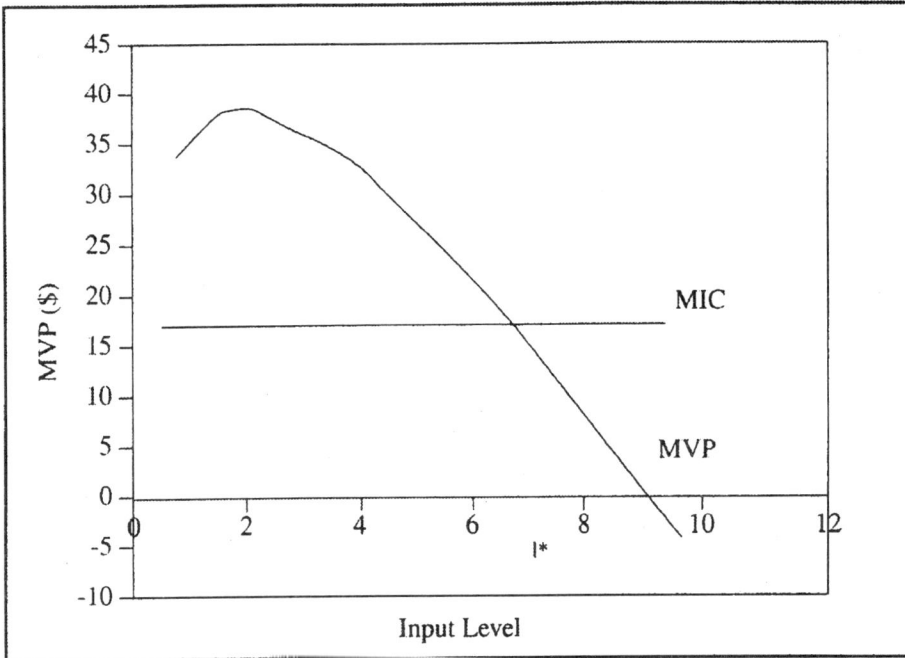

Figure 1. Optimal input use for a single variable input.

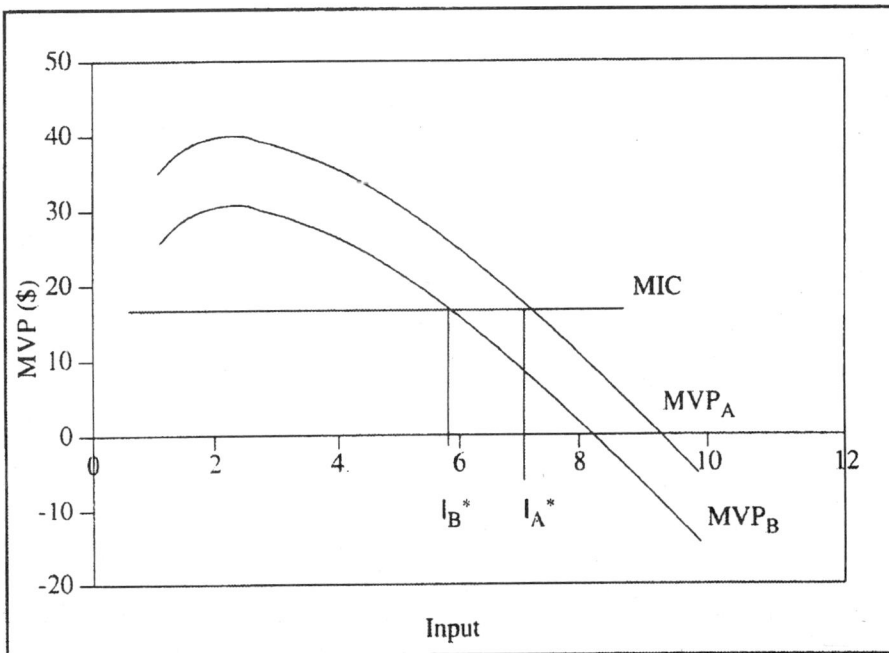

Figure 2. Optimal input allocation for soils of differing productivity.

decrease in yield on soil B due to the reduced input, and an increase in profits for the field. Note also that SSM results in a decrease in environmental damages because the over-application of input on soil B is eliminated. In the event that I_B^* is applied on all soils with a URT, SSM results in both an increase in input costs and an increase in yields on type A soils.

From the previous discussion, it is clear that both revenues and costs will be altered by SSM. Outlined in the following are the sources of costs and returns facing the farmer who adopts SSM. Throughout, it will be assumed that the decision rules used to manage the SSM System are primarily focused on the maximization of farm profitability.

Returns

In the absence of subsidies or other extra-market payments to the farmer, the only source of returns to the farmer for adoption of SSM technologies is in the value of the crop. Total gross receipts to the cropping enterprise is the product of crop yield, price, and the number of acres harvested. The specific farm situation to which the technology is applied and how the individual farmer chooses to manage the technology will determine the impact on each of these parameters.

Clearly, yield will be impacted by the application of SSM. However, it is difficult to argue the direction of change for this parameter. For instance, average yields could increase, decrease, or remain approximately constant (Table 1). It is almost certain that with application of an SSM that allows regulation of several inputs, yields on some sites (field locations) will increase while others will decrease. That is, SSM will allow identification of both areas of uneconomic over- and under-application of inputs. While SSM may reduce the variability of yields across the farm, it is not possible to predict the direction of change on average yields.

Price received clearly will impact gross receipts. Assuming that the farm was reasonably well managed prior to SSM adoption, crop quality for most crops is not likely to change sufficiently to impact price. However, there is a potential for increased commodity price with SSM. If crops are grown that have special characteristics (e.g., high lysine corn, organic crops, etc.) that will command premium prices, SSM allows improved ability to *preserve the identity* of these crops. Note, however, that this is not an automatic consequence of SSM, but rather it is an opportunity afforded to the adopter.

The number of acres harvested is also an important determinant of total receipts. Harvested acreage is likely to remain constant or decrease with SSM adoption. SSM technologies are management intensive. Provided that the farm manager makes many of the decisions and that the farm manager was fully employed prior to SSM adoption, he may be challenged to maintain the existing acreage. Over the next decade, new technological innovations likely will release this restraint on farm size. For example,

Table 1. Crop Enterprise Budget with Addition of Precision Farming Technology	
Returns: (price/yield/acreage)	
Yield	Constant, increase or decrease
Price	Constant or increasing
Acres	Constant or decrease
Total Returns	???
Variable Costs:	
Data costs (Grid sampling, mapping, remote sensing)	Increase
Fertilizer/lime material costs	Constant, increase or decrease
Fertilizer/lime application fees	Increase
Pesticide material costs	Constant, increase or decrease
Pesticide application fees	Increase
Total Variable Costs	???
Fixed costs:	
Depreciation	Increase
Interest on Investment	Increase
Development of management *human capital*	Increase
Total Fixed costs	Increase
Profit	???

if on-the-go soil tests and other forms of diagnostic/prescriptive remote sensing are developed, many of the analytical and decision processes can be automated, thus greatly reducing time management requirements of SSM.

The direction of change of farm total gross receipts is indeterminate (Table 1). It will depend on the relative increase or decrease in average yields, change in commodity prices, and change in enterprise size. These, in turn, will be influenced by site-specific factors. As scientists gain more experience with SSM, they will be better able to judge the relative contribution of each of these parameters.

Costs

There are two broad categories of cost that must be borne by the farmer. *Variable* costs are a function of the level of output of the farm. *Fixed* costs are invariant with changes in farm output. The adoption of a site-specific management system is expected to result in changes in both of these cost categories.

Variable Costs

Data Acquisition Costs. Site-specific management is an information intensive technology. As such, data acquisition costs will be substantial, at least for early forms of the technology. Geo-referenced soil sampling and scouting for weed, insect, and disease pests can represent sizeable production expenses. Data purchase, subscriptions, consulting fees and data management costs also can be significant. For instance, anecdotal evidence from conversations with industry personnel suggest grid sampling costs of $2.50 per acre (three acre grids), soil test lab fees of $5.50 per sample ($9.50 for samples with micronutrient evaluations), $3.00 per acre for geo-referenced field scouting, and $0.50 per acre for generation of yield maps.

There are a number of technological developments that could substantially lessen data costs. Remote sensing using satellite-based photography may be substantially less costly than field scouting and soil sampling; of course, reliable interpretation of such imagery must be developed. Even more promising would be the development of on-the-go sensors that would allow measures of soil fertility, identification of weeds, or other problems at the time of planting, spraying or other operation. Such developments would reduce the variable costs of data acquisition, labor, and management, but would increase capital investment and fixed costs. On-the-go sensors would also go a long way toward relinquishing the restraints on the farm size mentioned earlier.

Fertilizers. Fertilizers, including agricultural lime, represent a sizable cost item for crop producers. However, the direction of change for this cost category is not clear (table 1). For sites that have had a history of uneconomical over-application of nutrients or simply have a soil type that is rich in nutrients, grid sampling may reveal that fertilizer applications can be reduced. Other sites, perhaps within the same field or farm, may reveal that nutrient application rates should be increased from the uniform rates previously applied. Thus, the direction of change in fertilizer material costs will vary by site and circumstance.

Agricultural lime may be a case where input levels can be expected to diminish with variable rate application. Soil pH has been found to vary substantially within fields. Site-specific management of this input may result in zero application of lime over portions of the field, saving material costs and potentially reducing application costs. In fact, variable rate lime application has become the *posterboy* for the variable rate fertilizer industry. It is often the first service sold to farmers.

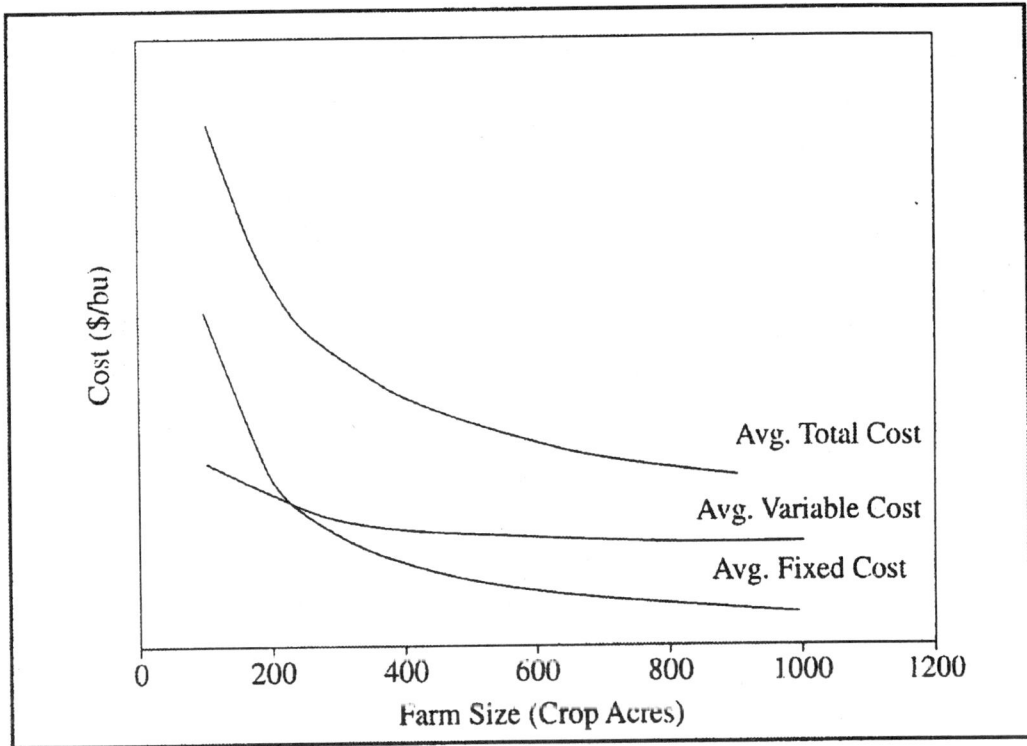

Figure 3. Expected average cost relationships for site-specific management technologies.

Fertilizer application costs will clearly increase with SSM (table 1). Anecdotal evidence suggests that service firms charge a premium of $2 or more for variable rate application of fertilizers relative to a uniform application. Even if the individual owns the variable rate application equipment, application costs are expected to rise due to increased labor costs associated with the application.

Pesticides/herbicides. Herbicides and pesticides also represent a major share of crop production costs. Although variable rate application of these inputs is still in its infancy, logic would suggest that there is potential for sizable economic return from SSM. Post-emergence herbicides, much like agricultural lime, may benefit from the site clustering of the input needs. If weeds are *patchy,* spot spraying may be an option so that at many sites herbicide applications will be zero. Gains may also be possible for pre-emergence herbicides. Research has suggested that different rates of herbicide application may be required on different soils. Thus, a variable rate application of these materials based on soil type and other parameters may allow cost savings. Although the potential may be ripe for herbicide/pesticide cost savings, the amount of these inputs applied will be a site-specific decision. On some sites, application rates may increase from the current (uniform) applications level. It Is also clear that the costs of scouting and variable rate application will rise relative to the uniform application strategy.

Other variable inputs. A number of other inputs can be regulated according to specific site conditions. Seed varieties and planting populations can be varied with soil type, slope, moisture conditions and other parameters provided that measurements for these parameters are available at planting time and that an algorithm exists with which to regulate the variety and rate selection. Because soils vary across the field, there may be substantial payoff to varying seeding rates and varieties with soil conditions across the field.

Labor and management are inputs that will not be regulated by the SSM technology, but rather are required inputs. SSM is an information technology. With the current state of the technology, human intervention in the decision process is quite important, and implies a substantial time commitment on the part of the manager. Thus, these input costs will increase with SSM adoption.

Fixed Costs

Fixed costs represent those inputs that are invariant with the level of production. Generally, these are annualized costs associated with durable capital investments. Examples of fixed costs of SSM are the depreciation, interest on investment, and insurance costs associated with yield monitors, computers and software, GPS equipment, VRT application equipment, and other necessary equipment.

Some farmers may choose to use financial leasing or custom hire in lieu of ownership for some durable assets. In the case of a financial lease, the farmer will face a fixed financial lease payment instead of a charge for depreciation and interest on invested capital. A farmer may elect to custom hire the variable rate application of fertilizers instead of owning VRT equipment. Generally, such operating leases are offered on a variable cost basis priced per acre or per day of operation. In this case, the full costs of VRT application are variable and appear in the fertilizer/lime/herbicides application fees in table 1.

Boundary mapping, soil and site surveys and the establishment of base maps also may represent a substantial expenditure. These, too, should be viewed as durable investments and their costs should be amortized as a fixed cost over a number of years. There is also a fixed cost associated with management. In particular, the cost of developing *human capital* is often omitted from the cost equation. Typically, there is a substantial learning process required prior to the effective use of the SSM system. The development of the knowledge base required to manage such a system may be substantial.

Profit

Profits are the difference in total receipts and total costs. The change in profits with the addition of SSM technologies cannot be determined in general—it depends on the circumstances for the specific farm. Total receipts will either rise or fall with SSM adoption for a particular farm. Similarly, variable costs can either rise or fall, depending on the magnitude of input savings (if any) realized with SSM. Only fixed costs are predictable, rising with SSM adoption. The change in profit will depend on the relative magnitude of changes in the cost and revenue categories.

Economies of Scale and SSM

Table 1 describes the expected sources of costs and revenues for any farm adopting SSM. However, the relative magnitude of these cost changes will not be the same for all farms. Farm size is an important modifier of average cost relationships. Figure 3 summarizes the expected relationship of per acre variable and fixed costs for farms of differing sizes.

Figure 3 is drawn with variable costs that diminish modestly with increased farm size. Although many of the variable inputs may be at a constant price regardless of quantity purchases, other inputs or services may include volume discounts. For instance, the variable rate fertilizer applicator may charge different rates per acre for 100 acres of application versus 1,000 acres.

The average fixed cost curve declines dramatically with increased farm size. If total fixed costs (e.g., the cost of owning the GPS, computer, mapping software, etc.) are equal for small and large farms, average (per acre) fixed costs are 10 times larger for the 100-acre farm than for the 1,000-acre farm.

The average total cost curve represents the total of average variable and average fixed costs. Clearly, the shape of the average total cost curve is determined primarily by the shape of the average fixed cost curve. The decline in average total costs with the expansion of farm size is referred to as economies of scale. Such economies are likely to exist for site-specific management technologies and will make adoption of this technology more profitable for larger farms rather than for smaller farms.

The Value of Information and Variable Rate Application

In this section, the value of SSM in the regulation of a single variable input is explored. It is demonstrated that site-specific information about local conditions (e.g., soil test data) may have value to the decision maker even if a uniform rate application is used. Variable rate application of the input may or may not provide additional economic value.

As an example, consider the case of a 21-acre field with seven equal-sized management units (grids). This field has variability of soil type and soil phosphorus levels that is not atypical of many real-world situations. The single input to be varied is phosphate fertilizer. Soil phosphorus test levels for each field unit are presented in Table 2 as are the P_2O_5 application recommendations. Soil phosphorus levels range from the 20 to 70 lbs/ac across the field. The average soil phosphorus level for the seven field units is 40.4. Recommended P_2O_5 fertilization rates range from 0 to 129 pounds per acre across the seven management units. The average of these seven fertilization rates is 57.9 pounds per acre. Also presented in table 2 are expected crop yields for each field unit without phosphorus fertilizers and for three alternative P_2O_5 fertilization regimes.

The base case represents corn production with the usual practice of a single soil test per field with uniform rate application of fertilizers. If three soil cores are taken from management units 2, 4, and 6, and mixed (averaged) prior to testing, the resulting soil phosphorus test result is 25.3 pounds per acre. This will result in a recommended P_2O_5 application of 93 pounds per acre. Notice that if, by chance, the soil cores are drawn from a different part of the field, eg. management units 1, 3, 5 and 7, the average soil phosphorus level is 51.8. The recommended P_2O_5 application rate for this sample is zero pounds per acre. Clearly, for a field with the variability of the example, the recommended application rates for a single soil test vary significantly depending on the location of the soil cores drawn for the sample. If the grid soil test results are treated as *truth,* the single soil test based on cores from grids 2, 4, and 6 result in an over-application of P_2O_5 (93 pounds for the single soil test versus an average of 57.9 pounds for the VRT method). Conversely, if the single soil test is based on cores from grids 1, 3, 5 and 7, phosphate fertilizer is underapplied by 57.9 pounds.

Expected corn yields also will vary dramatically under the three fertilizer application regimes. If no phosphate fertilizers are applied (the recommendation if the soil sample is drawn from grids 1, 3, 5, and 7), yields will vary from 132 to 219 bu/ac with a mean of 178.6 bu/ac (table 2). If 93 pounds of P_2O_5 are applied, the yield range is 188 to 230 bu/acre with a mean of 209.6. If the P_2O_5 recommendation is based on the average of the VRT phosphate rates (57.9 pound per acre) but the fertilizer material is applied at uniform rate across the field, the yield range is 173 to 230 bu/ac, with a mean of 202 bu/ac. The VRT application narrows the yield distribution to a range of 200 to 219 bu/ac, with a mean yield of 204.7. In this case, VRT substantially reduces the average application rate of phosphate fertilizer, and results in a somewhat larger average yield.

Variable input cost assumptions for the example are listed in table 3. Assume that the soil tests are drawn for general fertility decisions, instead of supporting the phosphorus fertilizer decision only. The single soil test will include both primary and micronutrient analyses at a cost of $9.50 per sample. For the grid sampling scenario, one sample will include micronutrient analysis and the remaining six only primary

Table 2. Soil Test Results and Recommended P2O5 Application Rates for Seven Field Management Units								
Field Unit (3.0 acres each)	**Average**							
	1	**2**	**3**	**4**	**5**	**6**	**7**	
Bray 1 Soil Phosphorus test (lb/ac)	70	20	62	2 5	30	31	45	40.4
Recommended pounds P_2O_5 per acre	0	129	0	106	82	77	11	57.9
Yield with VRT application of P_2O_5	219	200	214	200	200	200	200	204.7
Yield with 58 lb uniform rate of P_2O_5	230	173	230	184	193	194	210	202.0
Yield with 93 lb uniform rate of P_2O_5	230	188	230	196	203	204	216	209.6
Yield Without Additional Phosphorus	219	132	214	152	167	169	197	178.6

Table 3. Price Assumptions for the P_2O_5 Application Problem	
Cost of P_2O_5 Material Per Ton	$0.25
Cost of Soil Test (Primary Nutrients Only)	5.50
Cost of Single Soil Test (With Micronutrient Analysis)	9.50
Cost of VRT P_2O_5 Application/Acre	5.00
Cost of Uniform Rate P_2O_5 Application/Acre	3.00
Price of Corn ($/bu)	2.75

Table 4. P₂O₅ FERTILIZATION COSTS UNDER THREE APPLICATION REGIMES

	P₂O₅ Material	Soil Sample Costs[a]	Application Fees	Value of Yield Improve-ment[b]	Total
Single Sample, Uniform Rate Application:	$488	$1	$63	$597	$45
Sample cores from grids 2, 4 & 6					
Single Sample, Uniform Rate Application:	0	1	0	0	-1
Sample cores from grids 1, 3, 5 & 7					
Grid Sample, Uniform Rate Application	304	4	63	451	80
Grid Sample, Variable Rate Application	304	4	105	503	91

[a]*Soil sample costs are assumed to be divided equally among four fertility decisions (Nitrogen, Phosphorus, Potassium and Lime) and are amortized over 3 years.*
[b] *Yield improvement relative to no P₂O₅ fertilization.*

nutrient analysis at a cost of $5.50.[*] The costs of uniform rate and variable rate fertilizer applications are $3 and $5 per acre, respectively.

P₂O₅ application costs and returns for the 21-acre field under three fertilizer application regimes are presented in table 4. The base case scenario, a uniform rate of application based on a single soil test (grids 2, 4, and 6), results in a value of yield improvement from P₂O₅ (relative to zero P₂O₅) of $597. Although the soil sample costs are one-fourth those in a grid soil sample based regime, and P₂O₅ application fees are three-fifths those of a variable rate application regime, P₂O₅ material costs are much larger ($488). Net returns for the 21-acre field are $45 larger than if no soil test was made and no fertilizer material was applied.

The second scenario assumes that grid sampling is performed, but P₂O₅ is applied at a uniform rate based on the average of the seven fertilizer recommendations. This results in an application of 58 pounds per acre of P₂O₅ at a cost of $304 of material.

------

[*]The standard contract offered by one Central-Ohio service provider suggests that every sixth sample drawn will be analyzed for micronutrient content.

Although the application costs are the same as for the previous scenario, soil sampling costs are larger. The costs of this fertilizer application scenario total $371. The net return to the field, again relative to no fertilizer, is $80. The difference in the net return for uniform rate application based on the grid sampled data ($80) and the net return for the single soil test scenario ($1 or $45) is the value of information derived from the grid soil test ($35 or $81).

The final scenario considers the addition of variable-rate application of fertilizer. Soil test costs are the same as the previous scenario. However, fertilizer application rates will vary for each of the seven grids and application costs are higher. Total cost of the material is $304. Total cost of the application under this scenario is $413. Net return for the 21-acre field, relative to no application of fertilizers, is $91. Thus the variable rate application increases net returns by $11 for the 21-acre field relative to the uniform rate application based on grid sample information.

Environmental Consequences and Externalities

Agriculture, like other industries, has the potential to create pollution or other external costs to society. SSM has the potential to either increase or decrease the magnitude of agricultural externalities, depending on the relative increase or decrease in fertilizers, herbicides, and other agrichemical inputs. The profit-maximizing producer will not recognize these external costs. However, if policy-makers were to impose taxes or other mechanisms to charge (internalize) these costs to farmers, such taxes (subsidies) would become a private cost (benefit) and would be included in the individual farmer's profit calculation.

Examination of the hypothetical case presented in the previous section will give some insight as to changes with SSM that might impact the environment. The quantity of P_2O_5 applied varied substantially among treatment regimes (table 4). The two single sample, uniform rate application scenarios applied P_2O_5 fertilizer in a range of 0 to 93 pounds per acre. The grid sample, uniform rate application and VRT application resulted in the same total P application per acre (57.9 pounds per acre); however, this was allocated differently among the management units in the VRT regime. Soil P after application of P_2O_5 ranged from 29 to 79 pounds per acre across individual management units with the single sample, uniform rate app (base case).[*] The range of soil P for the grid sample, uniform app is 26 to 76 pounds per acre. For the VRT application, this range is only 33–70 pounds per acre. To the extent that environmental pollution hazard is correlated with soil P levels, environmental damages are lower with VRT versus URT application of fertilizer nutrients in the hypothetical case.

---

*One pound of P_2O_5 will raise the soil P test about 0.1 pounds (Ohio Cooperative Extension Service).

Summary and Conclusions

Site-specific management has the potential to improve both the profitability of the farm and to lessen environmental damages of agriculture. The farm level economic performance of SSM is site specific. Its financial performance will depend on the attributes of the farm's soils and other resources, the inherent variability in production for these resources, and previous management decisions. Yields likely will increase on some field sites and decrease on others relative to a uniform input application strategy. Likewise, the level of usage of fertilizers, pesticides and other inputs will vary unpredictably relative to URT. Farm total fixed costs are predicted to rise with SSM due to durable investments in machinery, mapping and resource inventories, and human capital. Profits associated with the SSM investment will be determined by the relative changes in revenues and costs.

The costs and profits of SSM will also be impacted by the size of the adopting farm. Economies of scale will be important for this technology, as it would be with any capital-embodied technology. Larger farmers will have a greater profit potential, and thus will predominate the early adopters of this technology. This may also mean that SSM, ultimately, will accelerate the trend toward larger, but fewer farms.

Environmental impacts also are likely. Because fertilizer and agrichemical input usage can either increase or decrease with SSM adoption, it is difficult to suggest how important this technology will be to reducing the aggregate environmental damages from agriculture.

Environmental costs and benefits are external to the farm firm. Farmers, when making the SSM adoption decision, will not necessarily consider these values. If it is determined that SSM has significant environmental value to society, adoption could be speeded by either transferring external costs back to non-adopter farmers through a tax mechanism, or rewarding adopters with a subsidy.

The above discussion is based on the assumption that the operator's primary goal is to maximize profits. An alternative would be to make decisions based on a joint function of profits and environmental benefits or, more simply, to maximize private profits subject to some environmental constraints on input usage. Such an implementation would use the power of SSM to directly impact environmental quality. However, using such rules would limit farmers' profit potential for the new technology. To encourage adoption of such environmental-friendly input allocation rules, transfer payments or tax credits could be granted for compliance.

Additional research is needed to understand both the private economics and societal issues. SSM will alter input usage at individual sites within a field. Average input usage may not vary greatly between SSM and uniform rate applications across the field. This raises the question of what is the environmental benefit of targeting fertilizers and pesticides, and effectively decreasing the variability of rates of application across the field and farm. It is likely that reduced variability of input usage will impact the environment's ability to assimilate agrichemical and fertilizer inputs. However, it is not clear how one would aggregate such site level impacts into an estimate of environmental benefits for a farm, watershed, or larger unit.

❏ References

Babcock, B.A. and G.R. Pautsch. 1998. "Moving from Uniform to Variable Fertilizer Rates on Iowa Corn: Effects on Rates and Returns." Journal of Agricultural and Resource Economics, Vol 23(2):385-400.

National Research Council. 1997. Precision Agriculture in the 21st Century: Geospatial and Information Technologies in Crop Management. Washington: National Academy Press. 149 pp.

Lang, L. 1992. GPS+GIS+Remote sensing: An Overview. Earth Observation Magazine. April 23-26.

Lowenberg-DeBoer, J. and S.M. Swinton. 1997. "Economics of Site-Specific Management in Agronomic Crops." In: The State of Site-Specific Management for Agricultural Systems, F.J. Pierce and E.J. Sadler, eds. Madison, WI: American Society of Agronomy. Pp. 369-396.

Ohio Cooperative Extension Service. 1985. The Ohio Agronomy Guide. The Ohio State University, Department of Agronomy. Bulletin 472.

Oriade, Caleb A., Robert P. King, Frank Forcella, and Jeffrey: Gunsolus. 1996. "A Bioeconomic Analysis of Site-Specific Management for Weed Control." Review of Agricultural Economics 18:523-535.

Swinton, S.M., and J. Lowenberg-DeBoer. 1998. "Evaluating the Profitability of Site-Specific Farming." Journal of Production Agriculture, Vol. 11(4):439-446.

Watkins, K.B., Y-c Lu, and W-y Huang. 1998. "Economics and Environmental Feasibility of Variable Rate Nitrogen Fertilizer Application with Carry-Over Effects." Journal of Agricultural and Resource Economics, Vol. 23(2):401-426.

CHECKUP FOR THE PLANET

Examining Environmental Health at the Turn of the Century

As we prepare to embark on the next millennium, people all over the planet are busily taking stock—reviewing the past, assessing the present and predicting the future. But how many are paying sufficient attention to health of the planet itself?

It's time to pause and assess the planet's condition. Like her human inhabitants, Mother Earth needs regular check-ups, and now is the ideal moment in history for a thorough exam. What follows is an overview of the Earth's environmental health record, focusing on Greenpeace's longstanding areas of greatest concern. Of course, all of these issues are interrelated, and all are vital to the well being of the planet and its complex ecosystems.

Forests

Condition. Rapidly Deteriorating

Symptoms. Loss of habitat, recreational opportunities, clean air and water; decline in species populations

Diagnosis. Half of the world's total forest cover has been destroyed, developed or converted for agricultural use. Only one fifth of the world's ancient forests remain in areas large enough to support their full range of native wildlife and ecological processes.

Industrial logging poses the greatest threat to the Earth's remaining ancient forests. Once cut down, the trees are converted into lumber as well as pulp and paper products, including toilet paper, newspapers, phone books and food additives. The United States alone accounts for nearly one third of the world's wood consumption.

Side effects. The livelihoods and cultures of millions of indigenous people are being destroyed.

The destruction of ancient forests further exacerbates global warming and changing weather patterns by releasing carbon into the atmosphere.

Prognosis. The remaining ancient forests are threatened by logging, development, mining, grazing and conversion to agriculture. An area of ancient forest approximately the size of one football field is lost every two seconds.

Treatment. Greenpeace is active on every continent where ancient forests still remain, as well as in the nations that are the largest consumers of ancient forest products—the United States, Japan and many European countries. Greenpeace has prioritized shifting corporate purchasing policies away from ancient forest destruction, setting aside large blocks of ancient forest as reserves, and halting the international trade of illegally logged wood products.

Toxics

Condition. Serious

Symptoms. Contamination from unseen poisons, impairing the health of humans and wildlife

Diagnosis. Some 170 organochlorine chemicals such as persistent organic pollutants, or POPS, have been found in human tissue and to a larger degree in animals. Over 11,000 organochlorines are currently in commerce. These toxic substances include DDT, dioxin and PCBs. Especially affected by this myriad of pesticides and other toxic substances are northern regions, particularly the Arctic, as POPs tend to migrate from warmer to colder climates. Native cultures suffer the highest exposure and the greatest risks.

The risks of POPs to human health include reproductive problems and cancer, with the greatest exposure for humans coming through the food supply. Of special concern are POPs passed on to children in the womb and through breastfeeding.

In addition, dioxin has recently been found responsible for the contamination of dairy products and meat in Belgium where animals were fed contaminated feed.

"We're bathed in PCBs, dioxin and various pesticides," according to Barry Commoner, ecological scientist and author, whose 1970 book *The Closing Circle* accurately predicted trends in energy and toxics for the next 20 years. "The petrochemical industry is just now acknowledging that they should have known about this toxic problem before they produced these chemicals and inflicted them upon us. But it's too late."

Side effects. Dramatic declines in the populations of European sea otters, as well as seabirds and birds of prey, have been linked to possible PCB exposure. POPs such as DDT and PCBs have been found in marine mammals in the Northeast Atlantic, indicating that toxic substances have contaminated the oceans and even the deep seas.

The toxic chemical dioxin, created and released when PVC is burned, travels long distances through the atmosphere, depositing on water and on land, including farm-land.

Prognosis. If the ongoing negotiations at the United Nations (UN) fail to produce an agreement to eliminate POPs and instead agree to merely "reduce" them, developing nations will be faced with increased levels of dioxin and other POPs. Practices such as open burning of trash containing PVC plastics and the use of unregulated incinera-tors will continue to release POPs into the atmosphere.

Treatment. POPs are a global problem; a global commitment is required to eliminate them. The United Nations Environment Program (UNEP) is attempting to negotiate an international treaty to eliminate 12 listed POPs and begin adding other POPs to the list. It is imperative that this effort succeeds. The nations of the world must also stand by their commitment to destroy stockpiles of banned or obsolete POPs and stop the generation of new ones. We must work to change outmoded pollution control laws that attempt to manage these inherently unmanageable poisons.

Climate

Condition. Critical

Symptoms. Rising temperatures, increased droughts, floods and wildfires, displace-ment of communities and ecosystems

Diagnosis. The burning of fossil fuels—oil, coal and gas—continues to accelerate global warming at alarming rates. Today, the symptoms are most noticeable in the Arctic, which is warming at three to five times the rate of the Earth as a whole. As pack ice melts or stays frozen for less of the year, the fragile Arctic wilderness is disrupted.

Globally, the devastation wrought by extreme weather has increased significantly. In 1998, the warmest year ever recorded, natural disasters such as ice storms, floods, wildfires, heat waves and droughts caused over $92 billion in damage, more than in the entire 1980s. The International Federation of Red Cross and Red Crescent Societies warns, "The explosive combination of human-driven climate change and rapidly changing socioeconomic conditions will set off chain reactions of devastation leading to super-disasters."

Side effects. Coral reefs are the rainforests of the oceans, homes to rich and diverse ecosystems and have an economic value in the hundreds of billions of dollars per year. Coral reefs are extremely temperature sensitive and the recent warming has put enormous stress on coral reefs around the world. In 1998, there was a worldwide outbreak of "coral bleaching," a sometimes deadly response of coral to these temperature changes. Reefs in many parts of the world suffered the death of more than 90 percent of live coral. If tropical waters continue to warm, scientists predict that bleaching events will increase until—by 2030 to 2070—mass coral death on the scale of the coral bleaching in 1998 occurs every year.

Prognosis. Greenpeace analysis of global warming science concludes that burning more than one-fourth of the oil, coal and gas currently in the world's economic reserves would greatly increase the risk of a disastrous climate shift. At the present rate of consumption, we will burn this amount within the next 40 years.

Treatment. "Fixing this problem is going to require a massive change in our systems of production, which are almost entirely based on burning fossil fuels," said Dr. Barry Commoner. "It would mean converting our entire electric and power energy industry into alternative sources."

Renewable energy sources such as wind and solar offer the best opportunity to replace fossil fuels without generating new problems. In fact, wind power is now the world's fastest growing energy source and is already cost competitive with fossil fuels. Solar panels can bring electricity almost anywhere, without the waste, pollution and complexity of conventional power plants. A simple commitment to build a large-scale solar factory could lower the price of solar panels to competitive levels.

Despite the benefits of wind and solar power, governments continue to encourage new fossil fuel development and open new frontiers to oil, gas and coal. While the major Federal incentive for renewable energy lapsed this summer, both Congress and the Clinton Administration have expanded support for the oil industry, which already receives $5 to $11 billion in subsidies a year.

Oceans

Condition. Serious, and steadily declining in many parts .of the world

Symptoms. Depleted fish stocks, reduced marine wildlife populations, damaged habitat

Diagnosis. Factory trawlers, which can net up to 500,000 pounds of marine life from the ocean per tow, represent today's largest threat to marine ecosystems. Fish catches are on the decline for the first time this century, as nearly 70 percent of the world's marine fish stocks are considered heavily exploited, overexploited, depleted or slowly recovering. Other species—like the Steller sea lions in the North Pacific Ocean—are in jeopardy because the fish they rely on for survival have been targeted by industrial fishing fleets.

Commercial interests also continue to haunt the word's whale population. Japan and Norway repeatedly ignore the hunting moratorium issued by the International Whaling Commission (IWC) in 1996. As a result, the number of whales killed commercially each year continues to increase.

"Water quality and fisheries are problems of intense concern in coastal areas," adds Dr. Thomas Malone, professor at the Horn Point Laboratory of the University of Maryland Center for Environmental Science. "We're making decisions on the ecological management of these areas based on very little knowledge and understanding. We simply are not totally sure how these areas are changing and how the current changes are affecting the quality of life in these food webs."

Side effects. In some studies, overfishing has been blamed for the loss of more than 40,000 jobs in eastern Canada and in U.S. fishing communities. As the human population and its dependency upon marine resources continues to increase along the world's coastlines, the damage left by industrial fishing fleets will have an increasingly profound effect on the quality of life on land.

Prognosis. Instead of holding owners of factory fishing vessels accountable for the mass environmental damage they cause, many nations' governments back their industrial fishing industries with generous subsidies. As long as this support continues, the decline in fish stocks and the destruction of habitats will continue.

Further, whale populations will diminish even more rapidly if Japan and Norway, with the help of Russia, are successful in lifting the ban on international trade with the lucrative Japanese market in whale products. Their efforts to reopen trade will occur at the Convention of International Trade in Endangered Species (CITES), in April 2000.

Treatment. Banning destructive industrial fishing methods is the best prescription for improving ocean health. In addition, industrial fishing must be regulated using a broader, more precautionary approach. "Too often, regulators look solely at the population of an individual species," said Niaz Dorry of Greenpeace. "It's imperative to examine the health of the habitat because it takes an ecosystem to raise a fish."

Moreover, global sanctuaries are needed to ensure that whale populations can thrive in an environment free of commercial hunting. Santuaries would also promote the emerging business of whale watching, a profitable alterative for those countries still engaged in hunting and trading.

Advancements in fisheries science cannot come fast enough. "Our biggest challenge," according to Malone, "is to develop the capabilities to observe changes in the health of coastal ecosystems and analyze that data in a timely fashion for developing sound fisheries management policies."

"Regulators are forced to make management decisions now, despite not having the information on the status of the fish stocks or on the impact of catching those stocks on other fish species and marine animals," added Dorry.

Genetic Engineering

Condition. Serious

Symptoms. Seventy million acres of genetically modified crops have been released without any long-term study of the potential threats to the environment. Soon we could start seeing symptoms such as the development of superweeds and superinsects, allergic reactions in humans and resistance to antibiotics.

Diagnosis. Genetic engineering poses an entirely new kind of pollution on the planet. It was just five years ago that biotechnology companies began selling seeds from genetically altered plants to farmers, releasing organisms that never existed before into the environment.

Corn, canola, soybeans, potatoes, tomatoes and strawberries are only some of the crops that biotech companies have tinkered with. Today, up to 60 percent of food in the United States contains genetically modified ingredients.

"The shocking thing," said Commoner, "is that many of the same companies that gave us PCBs, dioxin and pesticides without really understanding their impact on the world of life—companies like Monsanto, Novartis and DuPont—are now making the same error by manipulating genetics and food crops without really understanding what the consequences will be."

Side effects. The possible evolution and rampant spread of superweeds and superinsects—undesirable plants and insects with a high tolerance for pesticides—may pose grave dangers for the future. Farmers may be forced to use ever-increasing levels of chemical pesticides to prevent out-of-control plants from taking over fields and to keep invasive insects at bay. Pesticide runoff pollutes drinking water and waterways, endangering human health, killing fish and making streams and rivers hazardous for humans who may swim and fish there.

Prognosis. With genetic engineering we enter a strange new world. New organisms released into the environment have the potential to wreak havoc on fragile ecosystems and threaten human health in ways we are not yet aware of, because the technology is still so new. The outcomes could be biological pollution that is both widespread and irreversible.

In the process of killing destructive insects, plants genetically engineered to produce their own pesticides could kill beneficial insects too. Already, the Monarch butterfly appears to be a victim. A study revealed that pollen from genetically engineered corn could cause excessive death rates in Monarch caterpillars.

Patented gene stock controlled by biotechnology companies could lead to increased corporate hegemony over our food supply system. The agribusiness industry is pursuing the development of sterile seeds, which yield only one generation of crops.

Treatment. The United States must not allow further release of untested genetically modified crops, and instead pursue alternative agricultural strategies. The United States also must require labeling of all foods that contain genetically altered ingredients.

What is the best solution for consumers? Avoid processed foods, especially foods that contain soy, corn, cottonseed and canola, and support organic farmers and food companies.

A 21st Century Turnaround?

The threats facing the planet have been growing steadily over the last one hundred years. The various areas of concern are not just isolated problems, dealt with one at a time; they are intricately connected and could add up to disaster for the Earth.

What has led to these frightening changes to our planet? There are many answers, but the simplest is money. Large multinational corporations are on an unrelenting search for profit. These companies are not ruled by ethical and moral judgement, but by greed. Industrial processes such as PVC production, factory farming and fishing, logging, and oil exploration and development are directly linked to the deterioration of the environment. The United States government is also a major player in the destructive practices, doing everything from opposing an international treaty to eliminate POPs to offering huge subsidies to oil companies.

The most unnerving thing about all of this is that for many of the key environmental problems, there are solutions. Rampant clearcut logging can be replace by eco-friendly forestry, recycled products and non-wood materials. Clean renewable energy sources can allow us to phase out the dangerous carbon emissions from fossil fuels. Instead of driving fish and those species that depend on fish to survive to the edge of extinction, sustainable fishing practices can be implemented. A global commitment and subsequent action to eliminate POPs could save millions of lives. Organic agricultural solutions can be explored rather than experimenting on our food with genetic engineering.

We must all make a commitment to change our behaviors and the behavior of large, multinational corporations before it is too late. By working together, we may just be able to help Mother Earth recover from the environmental destruction she now endures.

MIND WHAT YOU EAT

André Voisin

The bountiful quantities of foods available in our grocery stores are often beautiful to look at, but how often do we consider the quality of the particular produce we have purchased? For the most part we assume that fruits and vegetables that look fresh and unblemished are healthy. But in this chapter from Soil, Grass & Cancer, André Voisin discusses the differences in food quality based on the quality of the soils in which the food is grown. He also discusses one of the saddest facts of modern agriculture, that plant quality is given attention when it affects the bottom line of commercial interests instead of disastrous consequences to human health.

The Appearance of a Fruit or Vegetable is No Indication of its True Quality. Vast propaganda campaigns have been undertaken the world over with the aim of improving quality of agricultural products. This generally means that perfection in size and appearance is demanded from a fruit, but not a thought is given to whether this fruit, so beautiful to look at, is deficient in one of the essential vitamins or mineral elements.

Study of Quality in Agricultural Products. At Geisenheim in Germany a "Federal Centre for Research into the Quality of Agricultural Products" has recently been set up. If only similar institutes and research centres could be established in every country and the quality of an agricultural product judged by standards other than its perfect external appearance or good measurements! The German research centre is under the direction of Professor Schuphan, some of whose work has already been dealt with, illustrating the influence of "dead" mineral matter on elements so fundamental to life itself that they bear the name vitamins. A further investigation by this great scientist into the quality of fruits will now be described.

Great Variation in the Vitamin C Content of Apples. Schuphan has shown, among other things, that the vitamin C content of apples can vary within wide limits. (Vitamin C or ascorbic acid plays an important part in cell oxido-reductions, in the integrity of vascular walls, in the growth and general resistance of the organism. Its

From *Soil, Grass and Cancer* by André Voisin. Copyright © 1999 by ACRES, U.S.A. Reprinted by permission

194

PROOF IN NUMBERS

Inventory numbers show that dairy and pork production is shifting to new ground. For example, the number of pigs in Oklahoma has increased, but the number in North Carolina has decreased.

Livestock Inventory by State (1,000 head)

	1995	1996	1997	1998	1999*
Hogs & Pigs					
IA	13,400	12,200	14,600	15,500	15,500
MN	4,950	4,850	5,700	6,000	5,500
NC	8,200	9,300	9,600	10,300	9,900
OK	1,000	1,320	1,650	1,980	2,340
UT	62	163	295	380	390
Milk Cows					
CA	1,257	1,268	1,314	1,429	1,476
IA	250	250	244	222	215
ID	240	265	279	296	324
MN	600	585	580	550	550
NM	196	195	205	219	235
WI	1,482	1,428	1,388	1,366	1,364

*1999 numbers are as of September USDA reports.

However, recent public pressure has tempered even Western expansion. For instance in Oklahoma, citizen concern over environmental issues recently caused the state to halt permit application for Seabord Corporation's proposed 27,000-sow Beaver County farm.

Many U.S.-owned pork integrators have purchased or built farms outside of the United States, in South America, where environmental concerns are less of an issue. Carroll's Foods (now Smithfield), has established large sow herds in Brazil and Mexico. Breeding companies also are expanding into South America. DanBred, a major European pig-genetics company that entered the U.S. in the early 1990s, recently established its first nucleus herd in Brazil.

Not just pork. Large pork farms are not alone in facing water quality pressures. Any concentrated animal feeding operation (CAFO), be it hogs, dairy, poultry or cattle feedlots, should be prepared for a future with stricter manure regulations.

"We were told by the Environmental Protection Agency (EPA) three years ago that the focus is going to be on agriculture," says Debbie Becker, director of the Washington State Dairy Association. "An electrifying message has caught fire with citizens that water is an essential life element and no one should screw it up."

This month marks the completion of the EPA rulemaking procedures for state-issued National Pollutant Discharge Elimination System (NPDES) permits for CAFOs. Aside from mandating waste-management plans and restricting when and where producers can spread manure as fertilizer, the permits may require some farms to replace lagoons if individual states find them below EPA standards. States are expected to begin issuing the new pollution permits in the spring of 2000.

"Producers need to understand that if they fall within the CAFO description and do not have an NPDES permit, they can be sued by citizens under the Clean Water Act," Becker says.

Dairy

In the last decade, milk production has moved West, with California surpassing Wisconsin as the No. 1 milk production state. Texas, Idaho and New Mexico now rank in the top dozen production states.

Water quality concerns, however, are putting the brakes on Western dairy growth. In California's Central Valley, home to 1,600 dairy farms, environmentalists contend they live in a cesspool where farm runoff has killed off the salmon population. And in Southern California, the Regional Water Quality Control Board gave Chino Basin dairy farmers a Dec. 31, 2001, deadline to remove a 2-million-ton manure stockpile blamed for polluting groundwater.

In Washington, water quality issues have grown along with dairy production. In 1998, an enterprise made up of 10 dairies near Sunnyside, Wash., was hit with a citizen lawsuit filed under the Clean Water Act claiming water pollution. Courts found one of the dairies liable for 15 violations of the act.

Environmental pressure has ushered in the return of slatted floors for dairy manure management. The waffle-style slat flooring allows manure to fall into pits, where it can be pumped for field incorporation. Most pits allow for six months of storage, addressing issues of manure containment and odor. The drawback is cost. A waffle floor for a 200-cow holding area costs about $40,000, and the entire system including pits and pumps is about $200,000 more upfront than lagoon or scraper systems.

As with hogs, arid states are welcoming dairies and their lagoons. North Dakota, as part of an alliance with South Dakota and Minnesota, is helping forage growers put together projects that include dairies. There are currently more than six projects with dairies of 800 to 1,200 cows.

Hogs

With approximately 4,000 active lagoons, North Carolina was the first state to incur a moratorium on hog expansion, extended until 2001. In March 1999, Gov. Jim Hunt announced plans to phase out lagoons within the next 10 years. Most recently, the Department of Environment and Natural Resources relaxed its post-flood rules for the hog industry, allowing farmers for a limited time to spray higher amounts of effluent and restock hog houses. In exchange, Smithfield Foods announced it will donate $15 million to research for manure management technology.

Data from experimental manure-management systems, like biofilters, which use bacteria to remove organic matter, show improvements in air and water quality. However, the cost of such alternative systems compared to lagoons remains substantial. "Of the technologies we have seen success with, in the best case the cost is about three times more than the cost of lagoons," says Mike Williams, director of the Animal and Poultry Waste Management Center at North Carolina State University.

On average, the cost to build a lagoon and purchase equipment for irrigation, factored in over five years, is about $1.30 per hog for a farm with 3,600- to 4,800-hog finishing space. The cost for alternative technologies like upflow biofilters and bioreactors is $3 to $4 per animal, Williams says.

Even producers in traditional hog states, like Iowa, are feeling environmental pressure. The state recently ruled that all confinement farms must have manure management plans in place and better lagoon designs.

Yet citizen groups continue to claim the state allows lagoons to contaminate groundwater. In August 1999, the citizen group Iowa Citizens for Community Improvement alleged that a hog lagoon in Humboldt County leaked manure. Although the state found no evidence of a leak, the pork producer was pressured into replacing the lagoon with concrete slurry storage.

FAST FORWARD

A VISIONARY LOOK AT THE NEXT 10 YEARS OF U.S. AGRICULTURE

Joe Roybal, Beef

"Predicting is tricky, especially about the future," baseball manager and philosopher Yogi Berra once said.

Berra's observation is an obvious truth. In less than one month, the new millennium will be upon us. What will it take to persevere as an agricultural producer 10 years from now?

The high-income farm and ranch in 2010 will be leaner and more efficient, generating a minimum of $250,000 in gross annual sales. The managers of those operations will be more market savvy, Internet proficient, consumer-oriented and willing to embrace technology and seek outside expertise. Most say they'll be buying inputs electronically.

That's the profile that emerges of the agriculturist determined to thrive through the year 2010, according to an exclusive survey of high-income farmers and ranchers.

Commissioned by the Agribusiness Division of Intertec Publishing, the survey polled readers of the division's nine agricultural trade magazines, representing every segment of U.S. agriculture and every region of the country. Polled were the readers of *BEEF Farm Industry News, Hay & Forage Grower, National Hog Farmer, Soybean Digest, Delta Farm Press, Southeast Farm Press, Southwest Farm Press* and *Western Farm Press.*

The survey asked respondents to give their thoughts on various questions regarding their individual—and their industry's—prospects for the year 2010.

Serious concerns, but determination. What emerged was the portrait of a majority of respondents anxious about:

Low prices. "I'm 72 years old and just sold wheat at my local co-op for $2.02/bu., the same price my first crop brought in 1948. That does not give much hope for the future of agriculture."

The cost of inputs. "We farmers are working for no pay as taxes and machinery go up."

Environmental regulations. "If the government continues to ignore sound science, the future of the livestock business is bleak."

WHICH OF THE FOLLOWING PRECISION FARMING TOOLS DO YOU USE?		
	currently	**in ten years**
remote sensing	2.4%	17.4%
grid soil sampling	19.5%	33.5%
variable-rate planting	11.3%	28.4%
variable-rate fertilizer	18.4%	45.4%
variable-rate pesticides	9.2%	26.1%
yield monitor with GPS	13.3%	44.0%
none	62.5%	39.4%

A shrinking labor supply. "There aren't enough young people coming back to the farm or ranch, and there's a lack of quality employees."

Competitive markets. "We are losing market share to countries that not only provide their producers with price supports, but export subsidies."

Concentration in agriculture. "The corporate world will soon control the food to fiber chain, from 'dirt to dinner.'"

But, at the same time, many respondents expressed a dogged determination and an aggressive, proactive desire to remain competitive and viable in the next century.

"Our future is bright if we learn to negotiate the value of retail products back to the farm," wrote one respondent. Another said, "Americans are very innovative when forced to be—they will meet the challenges of the future." Still another said: "The future is extremely bright for good managers of people and finances."

Building more efficiency, productivity and marketing proficiency were respondents' most mentioned strategies to guarantee survival. Eighty-two percent think $250,000 or more in gross annual sales will be needed to be competitive in 2010. Sixty percent believe that threshold is at least $500,000 in annual gross sales. That response differed by age of operator (see chart) and size of operation.

The median age of respondents was 52 years. Of respondents, 93% grew crops, while 48% were commercial or seedstock beef producers and 19% sold fed cattle. Another 14% sold slaughter hogs.

Those surveyed were also asked questions specific to their agricultural segment, On the following pages of this special issue sponsored by mPower[3], the editors of Intertec's nine Agribusiness Division magazines go into detail on specific industries.

Expansion is part of the picture. Less than 40% of row crop respondents expect to expand their acreage. But those who do plan to expand expect to add an average of

Use of the Internet on farms	
$100,000–249,999	25.3%
$250,000–499,999	38.5%
$500,000+	50.8%
Variations in Internet use based on age of farmer	
Under 35	62.5%
35 to 54	45.7%
55+	20.7%

70% more acres. And that expansion will be accomplished, they say, primarily by renting (70%) and purchasing (61%) land.

This expansion in tillable acres is favored more by agriculturists with $500,000 in gross annual revenue (48%) than by those below $500,000 in gross annual revenue (about 30%).

Most respondents commented that technology and outside expertise will be necessary for American producers to compete in the coming decade. Cropping respondents expect to see significant jumps in the adoption of remote sensing, grid soil sampling, variable-rate planting, variable-rate fertilizer, variable-rate pesticides and yield monitoring with global positioning.

Which of the following technologies or services do you use in your beef operation?		
	currently	in ten years
sexed semen	3.0%	18.3%
electronic ID	3.7%	26.2%
embryo transfer	8.5%	19.8%
value-based marketing	18.9%	33.3%
artificial insemination	34.8%	46.8%
beef marketing alliances	22.6%	50.0%

How much do you need to gross annually to continue farming in 10 years?

■ under 35 years of age ▨ 35 to 54 years of age ▨ 55⁺ years of age

(%) percent reporting (y-axis: 0, 5, 10, 15, 20, 25, 30, 35, 40, 45)

x-axis categories: under $249,000 | $250,000–$499,999 | $500,000–$999,999 | $1 million⁺

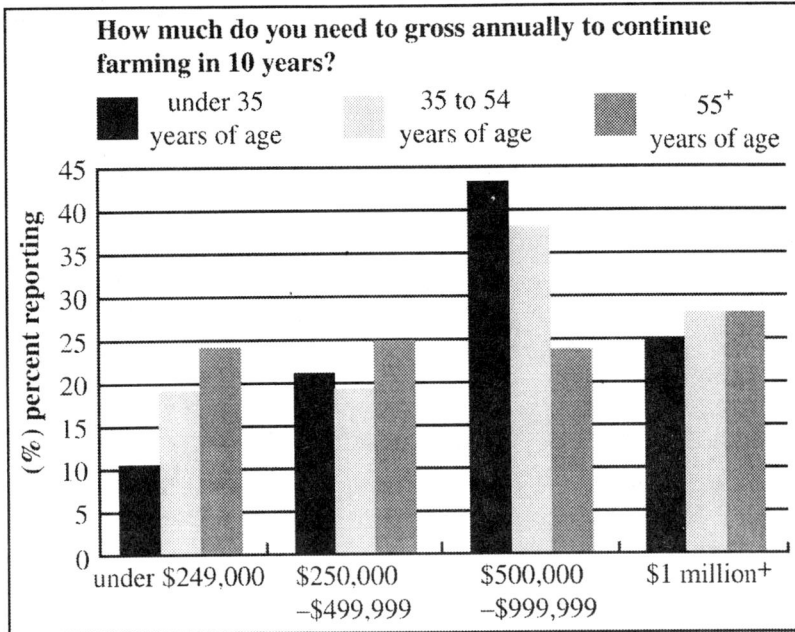

On the livestock side, responding cattle producers expect to see significant jumps in the use of sexed semen, electronic identification, embryo transfer and artificial insemination.

Significant growth and change are forecast for the Internet. Of the 36% of respondents who said they currently use the Internet, 78% said they use it primarily for e-mail.

Only 6% of respondents had actually purchased inputs on the Internet, but 48% indicated their intention to do so over the next decade. The most likely inputs to be bought over the Internet were chemicals, equipment, parts and seed.

As a rule, the segment tending to be most aggressive in expansion, adoption of technology and overall optimism was the younger producers (less than 35) and operations generating more than $500,000 in annual gross revenue.

But what do economists say? Is this an accurate picture of U.S. agriculture in the year 2010? Richard Brock, economist with Brock Associates, Milwaukee, WI, believes so.

"Particularly in the next two years, we'll go through one of the more dramatic changes that we've seen in the last 20 in row crop production—simply due to the fact that the average age of these farmers is very high.

"Largely, this segment is unable to adapt to the Freedom To Farm rules, and is taking early retirement. So we'll see a loss of about 15% of these people to retirement, not forced liquidation," Brock says.

This, he says, will accelerate the size of row crop production units appreciably in the very near term, with the average size of farms between now and 2010 growing to three times today's average size.

Iowa State University livestock economist John Lawrence expects to see "dramatic" changes in both the pork and beef industries in the coming decade.

In the hog industry those early adopters are the operations that positioned themselves for the current downturn, Lawrence says. When the dust settles, they'll be the survivors.

On the cattle side, Lawrence says the heightened beef demand that has occurred since July 1999 could portend better times for beef producers.

MAJOR SHIFTS IN OWNERSHIP

Dale Miller

Pork producers hope the 21st century will bring significantly stronger hog markets than they're seeing at the end of this century.

Whatever optimism they can muster—and there is some—is of the cautious kind.

Industry stakeholders move into the new millennium knowing that the term "big" is being redefined. Buyout notifications and merger announcements in the waning months of 1999 signaled significant shifts in ownership.

Capitol Hill lawmakers are increasingly being pressured to contain this momentum of the big getting bigger. Add to that the calls for mandatory price reporting and antitrust litigation, and it becomes even more challenging to make solid predictions about the first decade of the new century.

Smithfield Foods, the world's largest pork packer, shook the industry this year by starting to make good on its stated intent to capture the production and processing efficiencies of vertical integration.

It began 1999 by purchasing Carroll's Foods, J & K Farms and Western Pork Production Corporation. Apparently, that was just a start. In September it announced a purchase agreement with the nation's largest pork producer, Murphy Family Farms, followed within weeks by the acquisition of Tyson Foods' swine operations—bringing its total sow count to a whopping 785,000.

Mix in the unusually low prices pork producers have suffered the past 18 months, which slashed their income by 40–60%, and it's not hard to understand the predominant attitude revealed in our Intertec Publishing survey: a dampened optimism for the future and for expansion plans. (The Murphy and Tyson acquisitions came after our survey was conducted.)

Here's an analysis of what producers told us:

Producer profile. It's clear that age and income tempered attitudes about the industry. Over three-fourths are between 35 and 64 years of age, the heart of the industry, and likely are not newcomers.

The profile gets a bit murkier, however, when gross sales are factored in. Not only does distribution of income widen, but producers' marketing methods in recent years could have had serious economic ramifications.

Overall, just 19% told us they are selling hogs on a packer contract—below some industry reports. It is helpful to understand that 28% say they sold feeder pigs in the 1998 reporting year, and therefore would not have had packer contracts; 79% say they

---

HOW LIMITING ARE THE FOLLOWING FACTORS TO YOUR FUTURE GROWTH IN PORK PRODUCTION?	
market price	5.61
packer slaughter capacity	4.47
pork (meat) quality	4.35
foreign trade/exports	4.27
feed grain prices	4.24
market reporting/price discovery	4.22
herd health concerns	4.17
water quality	3.98
food safety	3.92
odor control	3.75
animal rights/animal welfare restrictions	3.34
employee availability	3.15
other	2.50

Note: 1 = least important, 6 = most important

sold finished hogs for slaughter; and just over 21% sold some breeding stock. Some sold pigs in all three categories.

Of those selling on packer contracts, one-third say their contract is a formula-pricing program tied to current cash markets, while 25% sell on a cost-plus (matrix) contract and an additional 25% sell at a fixed price tied to a futures contract.

When gross sales were considered, the number of producers with packer contracts edged upward. Of those reporting gross sales of $100,000–249,999 (low) and $250,000–499,999 (medium), 12% have packer contracts. But for producers with $500,000 or more (upper) in gross sales, twice as many (25%) have packer contracts. Gross sales figures were not restricted to respondents raising hogs only; some had other agricultural enterprises.

Checkoff's chances. A referendum on the mandatory pork checkoff is likely sometime in 2000. Of producers responding to a question about whether they support the checkoff, 65% answered "yes."

In the under-35 group, 56% favored continuing the mandatory program; in the 35–54 group, 68% were in favor. But those 55 and older were less supportive—only 46%.

What to expect in 2025

Maynard Hogberg, Michigan State University and John Lawrence, Iowa State University, peer into their crystal balls to see what changes the hog industry might face in the next 25 years:

1. Technology will continue to push costs lower and quality higher. Successful farms will quickly adopt new technology. Specific genes that impact feed efficiency, growth rate, reproduction, carcass quality, lactation, etc., will be identified and exploited.
2. All aspects of the industry will be more specialized—from breeding and genetics, all the way to the retail counter.
   Animal identification will be refined. Consumers will know how and where their pork was produced. "This will have many implications on meat quality, food safety and animal well-being," says Hogberg.
3. Margins to pork production will stabilize at market rates of returns for capital, labor and management. Producers will earn an opportunity wage rate, managers will earn a competitive salary and owner equity will earn a risk adjusted market rate.
4. Cloning will become more prevalent. Pig organs and tissue will be used more for transplants into humans.
5. Production systems will be low-odor, environmentally benign systems that can be placed in any agricultural setting.

"With change comes opportunity," believes Hogberg. "The challenge is to stay knowledgeable enough so that you're in front of the wave of changes and not behind them."

Using the three gross sales levels described above, 64% of the low-income producers, 53% of the medium-income group and 78% of the upper-income group support the checkoff.

Value-added corn. Asked if they currently feed value-added corn (high-oil, high-lysine or low-phytate) varieties, 21% say they do. Of those, 77% use high-oil corn, 15% feed high-lysine and 15% feed low-phytate varieties. Of the 79% not feeding value-added grains, 46% say they anticipate using them within the next 10 years.

Long-term expansion plans. Twenty-one percent of survey respondents say their long-term business plans call for expanding their hog operations. Asked to estimate an increase in numbers sold for the next five years, they believe their herds will grow by about 34%.

As expected, younger producers are more inclined to make expansion plans. About one-third of those under 35 years of age have expansion plans, but only 17% in the mid-range (35–54) and 19% in the 55 and older group want to expand.

When the differentiating factor is gross sales, 19% of the low-income group and 13% of the middle-income group have expansion plans, while 32% of the $500,000-plus group expect to expand.

Look to the future. Producers were asked to rank the importance of a dozen factors that could potentially limit their production plans. The chart shows the relative ranking, with the top two spots reflecting concerns over low market prices and tight slaughter capacities, which have contributed to the lowest hog market prices since the Great Depression.

Nearly 58% listed inadequate income as the factor most likely to cause them to quit farming. Retirement ranked second (41%), followed by government regulations (30%) and lack of financing (17%).

Of those responding to the survey, 86% expect to be farming five years from now and 74% expect to be farming 10 years from now.

Since our survey coincided with slumping prices for most agricultural commodities, we asked producers to temporarily put aside the current economic situation and classify their views of farming's future. The verdict was virtually split, with half describing their futures as "bright" or "promising," while the other half described it as "not so good" or "bleak."

Know thy competitors. When asked to name U.S. pork's biggest competitor in the export market, about one-third named Canada.

Asked which country would register the greatest pork production growth in the next 10 years, China, Brazil and Canada led the pack.

Ranch Setting Teaches Old-Fashioned Values, Discipline and Work Ethic

North Dakota Stockman

Home On The Range celebrates 50 years of helping troubled youth

In a legendary folk song, home on the range is where the deer and the antelope play. For 50 years, in Sentinel Butte, N.D., Home On The Range (HOTR) is where troubled youth have been given a second chance through personalized therapy and by learning the values that ranch life teaches.

Today, the residential child care facility, just off I-94, is home to 10- to 18-year-olds struggling with things such as alcohol and drug addictions, family problems, social skill troubles and many types of abuse. HOTR strives to teach the youth the skills necessary to be responsible, productive citizens by helping them address those problems and by developing their self worth, Kristie J. Asay, HOTR development director, said.

And what better place to nurture kids and teach them a hard work ethic and strong moral and spiritual values than a ranch? Those were the thoughts of Father Elwood Cassedy, a New Jersey-born Catholic priest, who first established HOTR in 1950. Cassedy had seen firsthand the troubles induced by neglect and homelessness as he worked in New York City in the 1920s. He watched boys, many near his own age, living on the streets and fighting, stealing and gambling just to survive. He made it his dream then to someday, somehow, provide a safe and loving place where neglected and homeless boys could thrive.

Cassedy's dream took shape following a speech he made to the Fraternal Order of Eagles in Deadwood, S.D., in 1949 in which he described the problems of youth and his vision to help correct them. The Eagles were so moved by Cassedy's talk that they "passed the hat" and came up with $123 that day to help get him started. Publicity of the Eagles' meeting generated even a more lucrative gift from Edward and Emma Lievens, an elderly couple from Sentinel Butte, who deeded HOTR's original 960 acres of farm- and pasture land for Cassedy's effort.

Shortly following, a wooden granary was remodeled and the first three boys came to stay with Cassedy in July 1950.

Cassedy knew from the beginning that the ranch would serve as an integral part of residents' progress, but he had no livestock and was a city boy with no farming

---

experience. That's when Ray Schnell, Sr., stepped forward as an agricultural advisor for the ranch and organized a benefit auction of donated animals. Donations of cash, livestock and other goods flooded in from church and civic groups, cattle ranchers and other ardent supporters.

It didn't take long for word to spread about HOTR. From the beginning, Cassedy received more requests for placements than he could accept. Increased and improved facilities allowed the center to gradually grow in staff and residents to a present size of 75 employees and 79 youth, including 20 girls, who were allowed into HOTR's program in 1990. Father William J. Fahnlander took the reins as HOTR superintendent after Cassedy's death in 1959.

Over the years, HOTR has produced a variety of animals, including beef and dairy cattle, horses, hogs, sheep, chickens and rabbits. Today, Loren Szudera, HOTR's farm stock manager, maintains the center's herd of 85 Gelbvieh and Gelbvieh-cross females and 19 horses and its hay production.

HOTR was home to beef cattle in its early years, but the enterprise was phased out and replaced with a dairy from 1970 to 1985. The high costs involved in maintaining dairy equipment and the early-morning labor strain on the residents were reasons ranch officials decided to switch back to a high-quality beef herd.

In 1987, the Fraternal Order of Eagles generously stepped forward once again, pledging to provide HOTR with 27 beef heifers, one for each individual Aeries in North Dakota and South Dakota, and two bulls from the ladies Auxiliaries to get them started in beef production.

Ray Schnell, Jr., also a HOTR supporter; had visited an area in Germany where Gelbvieh cattle originated. He was especially impressed with the uniformity of the cattle and also their fertility, milking ability and feed conversion and, so, recommended the breed for HOTR. The foundation Gelbvieh females were purchased, and several of the originals are still working cows in the herd today

Szudera said Gelbvieh have worked well at HOTR because of the cattle's mild dispositions. "When you have young people working with them, that is an absolute must."

And the residents do take a part in the cow-calf operation, often assisting in feeding, calving and branding. When a student arrives, he or she selects a job, whether it be feeding calves, preparing meals or something else, and some of the most popular jobs involve working with the animals, Szudera said, especially during calving, which begins about March 1. He said, "A lot of the residents have never seen anything born, and when a baby calf is born, that's when reality sets in. There's something about new calves and spring that [launches] a new beginning for them."

The cattle also provide a great way to teach responsibility, Szudera said. "The residents really take pride in taking care of the animals, and they do a great job at it. This kind of work has made a lot of positive changes for a lot of kids." And the HOTR staff lets the kids know that they're doing a good job, because, Szudera said, "If you want to see a kid melt like an ice cream cone on a 90-degree day, give him effective praise."

He remembers a frigid snowstorm one April. Szudera, who lives a few miles away from the HOTR site, anxiously hopped in his pickup and went to bring the cows up to the barns for protection from the winter blast. But when he got to the pasture, there were no cows to be found. To his pleasant surprise, a resident, concerned about the safety of the animals, had gathered them up himself and locked them away in safety.

The HOTR cows and heifers are mated to Gelbvieh bulls and Angus bulls, respectively, for a 60-day calving period. The calves are placed on a strict preconditioning and vaccination program and marketed following weaning through a local sales barn, with the steers averaging approximately 675 pounds. About 10 to 12 heifers are retained each year to develop as herd replacements. Szudera said he has always appreciated the marketing advice HOTR has received from the Schnell family.

One of Szudera's short-term goals is to incorporate a synchronized artificial insemination program to shorten the calving period and add more uniformity to the HOTR's calf crop. He believes this will also help HOTR improve its genetics at a quicker pace and teach the residents a new aspect of the beef business.

Asay said HOTR's cattle herd is something people identify with the child care facility and have for the last 50 years. "It's part of what we are and the way we are able to help kids," she said.

Identifying the Gap between Consumer and Farmer

Agri Marketing Editors

For ag and food companies to move further into the value-chain they need to understand the end-consumer. And it appears there is a lot to learn, or at least a lot of work that needs to be done, for farmer perceptions of consumers, and actual consumer perceptions to match.

Consumers often perceive issues surrounding food production differently than farmers, according to recent research conducted by Roper Starch Worldwide, and funded by American Farm Bureau Federation and Philip Morris family of companies. The research shows sizable gaps between consumer perceptions and how farmers perceive consumer perceptions.

The research, conducted in July and August 1999, sought to pinpoint consumers' views and expectations of the food supply and modern farming practices, as well as to learn farmers' perspectives on farming practices and meeting consumer demands. Roper Starch Worldwide interviewed 1,002 consumers with shared or primary food-shopping responsibility for their household, and 704 farmers who served as the primary decision-maker for the farm or ranch.

Farmers Underestimate Food Issues

Although farmers fairly accurately rate the importance and satisfaction of several food issues to consumers, they fall short in two areas: foods being grown without pesticides and meats and poultry not given antibiotics while being raised.

The mean importance on a scale of 1 to 10 of pesticide-free food for consumers was 7.9 while farmers thought it would be 6.1. Similarly, the mean importance for antibiotic-free meat for consumers was 7.5 while farmers predicted it, too, would be 6.1.

Other points of disparities in consumer and farmer perceptions lie in nutrition, chemicals, taste, and prices.

How Much, If At All, Do You Worry That the Foods You Buy ight Not Be Safe to Eat?

	A great Deal	Some	Only a Little	Not At All
Consumers worry	23%	38%		
Farmers personally worry	32%	34%		
Farmers think consumers worry	46%	12%		

❏ Nothing is more important than nutrition, say 86 percent of the consumers. Farmers dramatically underestimate this, with only 58 percent of farmers believing nutrition is the most important thing to consumers.

❏ For 74 percent of consumers, any price is worth knowing that food has no pesticides or chemicals. Meanwhile, 47 percent of farmers think that consumers would pay any price to be sure that food has no pesticides or chemicals.

❏ 53 percent of consumers say spots or size do not matter—"all I care about is knowing fruits and vegetables are free of pesticides and chemicals." Only 32 percent of farmers think consumers feel this way.

❏ Taste is most important to 68 percent of consumers. Here farmers overestimate consumer views—79 percent of farmers think the most important thing to consumers is taste.

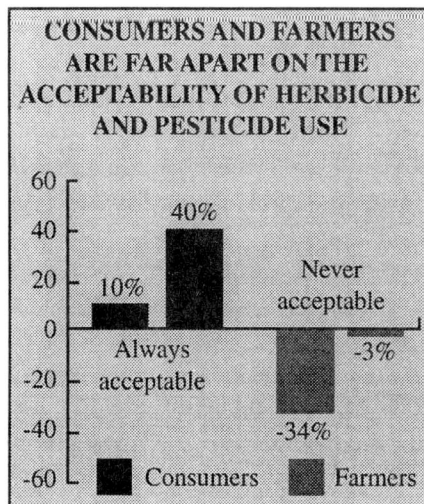

CONSUMERS AND FARMERS ARE FAR APART ON THE ACCEPTABILITY OF HERBICIDE AND PESTICIDE USE

❏ When choosing among several options, 44 percent of consumers buy the lowest priced foods available. Meanwhile, 56 percent of farmers think consumers look for low price when choosing what foods to buy.

Food Safety Bigger Concern than Expected

One of the largest gaps the research identified is in the area of food safety. Knowing the high standards in place, farmers do not worry a lot about the safety of the food supply. The research shows that two-thirds of the farmer respondents say they worry only a little or not at all about food safety.

And farmers believe consumer perceptions are similar to their own with nearly six in ten farmers responding that they believe consumers worry only a little or not at all about food safety. However, the reality is quite different. More than 60 percent of the consumers surveyed worry some or a great deal about food safety.

Food Production Perception Differences

In addition to food safety, consumers and farmers rate food production practices differently. Notable gaps exist between ways consumers and farmers believe food production can increase.

For example, 65 percent of the farmers believe more use of pesticides and herbicides would be beneficial to increasing food production. Only 25 percent of consumers share those beliefs, while 63 percent of consumers say such products are only harmful, not beneficial.

Similarly, 73 percent of farmers say more use of chemical fertilizers would be beneficial to increasing production while only 30 percent of consumers agree. Fifty-eight percent of consumers believe increased use of chemical fertilizers would be harmful to increasing production.

This relates to consumer opinions about water quality. Consumers are more than twice as likely as farmers to think that agricultural chemicals in the water supply present a major problem. Sixty-eight percent of consumers believe pesticides, herbicides and fertilizers from farms entering ground water is a major problem. Only 29 percent of farmers share those concerns.

Biotechnology Shows Promise

To produce enough food for a growing population, consumers overwhelmingly favor biotechnology and land preservation over the use of chemical fertilizers or pesticides. In fact, 73 percent of consumers, when asked about alternatives to the uses

WILLINGNESS TO ACCEPT THE FOLLOWING CONDITIONS AS TRADEOFFS FOR NOT USING CHEMICALS IN FOOD PRODUCTION		
	Consumer Willingness	Farmers' Views of Consumer Willingness
Biotechnology	73%	66%
Seasonal Availability	72%	28%
Smaller Selection	68%	26%
Pest Damage	52%	28%
Higher Prices	57%	29%

of farm chemicals, say they would be willing to accept biotechnology to develop crops that require fewer pesticides.

A large number of consumers say biotechnology is acceptable, with few consumers saying that biotechnology is "never acceptable." Generally, farmers find biotechnology or genetic engineering more acceptable than consumers do.

Biotechnology and genetic engineering rated highest among all alternatives to continued use of chemicals in food production. Options included: higher prices; food that is healthy but somewhat damaged by pests; smaller selection of foods; and seasonal availability of foods.

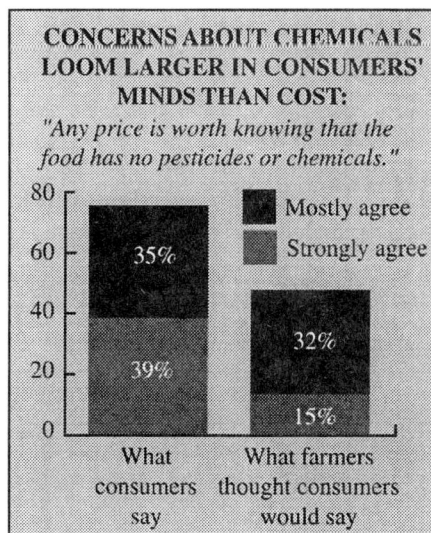

CONCERNS ABOUT CHEMICALS LOOM LARGER IN CONSUMERS' MINDS THAN COST:
"Any price is worth knowing that the food has no pesticides or chemicals."

While 37 percent of consumers say they've heard more about the benefits than drawbacks of biotechnology, their support of biotechnology increases to:

❏ 57 percent if biotechnology improves the taste of foods,
❏ 65 percent if biotechnology improves the nutritional value of food,
❏ 69 percent if biotechnology increases food production,
❏ 73 percent if biotechnology reduces pesticide use.

Farmers believe that consumers have heard more about the drawbacks than benefits of biotechnology. In reality, most consumers haven't heard much at all, good or bad.

Consumers "Don't Know"

Farmers and consumers agree that the agriculture industry could do a much better job of explaining the benefits and drawbacks of modern agricultural practices. Seventy-one percent of farmers and 67 percent of consumers agree that the agricultural industry is doing only a "fair" or "poor" job of explaining the benefits and drawbacks of farming techniques to the public.

"I don't know," and "I haven't heard about it," are the answers of choice for consumers when asked whether they have heard more about the benefits or drawbacks of seven common farming practices. These practices included irradiation (41 percent answered "don't know"), biotechnology (41 percent), antibiotics (31 percent), hormones (28 percent), organic production methods (28 percent) and pesticides and herbicides (13 percent).

The bottom line is that agriculture has an opportunity to educate consumers about the benefits of new agricultural technologies, especially irradiation, biotechnology, and the use of antibiotics to treat animal diseases. But such work must be done quickly.

The study shows that consumers are more likely to label the most widely publicized farm practices as "never acceptable" rather than "always acceptable." This is an indication that as more negative information is publicized, more people will find farming practices unacceptable.

Family Farm Inc.

Douglas W. Allen and Dean Lueck

---

*Commentators as different as Dan Glickman, Willie Nelson, and
Jane Smiley have recently voiced concerns that the industrialization of agriculture is consuming family farms, turning
rural landscapes into industrial parks, and forcing farm-folk
to become wage laborers. The data belie this concern: family
farms still dominate agriculture.*

This fact is indisputable even though farm numbers have declined, farm size has increased, and technological changes have converted farms into capital-intensive enterprises. Much of the concern implicitly assumes that changes in farm size and capital intensity have also led to changes in organization, and it assumes that what is happening in one farm sector must be happening throughout agriculture. The 1997 Census of Agriculture shows that more than 86 percent of farms are organized as "family farms." Excluding small family-held corporations, farm corporations made up only 0.4 percent of all farms in 1997. These corporate farms controlled just 1.3 percent of all farm acreage, and generated only 9 percent of all sales receipts. By contrast, the 1992 Economic Census shows that, outside of the farm sector, corporations generate more than 75 percent of all receipts. Farming is unique in the modern economy because of the relative unimportance of the corporate form of business organization.

Ronald Coase and Mother Nature

Why does agriculture continue to be dominated by family-based firms? The seeds of the answer lie in a framework developed by Ronald Coase in his work on the theory of the firm. Coase examined the tradeoff between incentives arising within firms and incentives stemming from the market. Because farms operate in unique circumstances defined by nature, understanding farm organization requires relating the modern explanations of the firm to special constraints nature places on growing food and fiber.

The main feature that distinguishes farm organization from "industrial" organization is its seasonality. Classical economist John Stuart Mill (1806–1873) said long ago that agriculture "is not susceptible of so great a division of occupations as many branches of manufactures, because its different operations cannot possibly be simultaneous. One man cannot be always ploughing, another sowing, and another reaping."

To the farmer, a "season" is a distinct period of the year during which a given activity (such as planting and harvesting) is optimally undertaken. For example, for spring wheat grown on the northern Great Plains, the month-long planting season usually begins in April, and the harvest season is primarily restricted to August. Seasonality influences the number of times per year the production cycle can be completed, the number of stages in the cycle, the total number of tasks (specific jobs) in a given stage, and the length of the stages. Nature's random forces also distinguish agriculture. Random events are particularly acute in many types of farming where weather and pests may strike unexpectedly. Thus, nature plays two distinct roles in farming: it governs the predictable seasonality of activities, and it strikes unexpectedly.

The key to understanding farm organization is appreciating the role nature plays in generating the incentives that favor family farms. First, random production shocks from nature generate opportunities for hired farm workers to shirk their duties. Second, seasonality limits the potential gains from specialization and creates timing problems between stages of production.

Family Farms vs. Larger Organizations

Farm organization can vary from a single owner-operator to a public corporation with many owners and specialized wage labor. A "pure" family farm is the simplest case: a single farmer owns the output and controls all farm assets. Factory-style corporate agriculture is the most complicated case: many people own the farm, and labor is provided by large groups of specialized wage laborers. Partnerships are intermediate forms in which a small number of co-owners share output and capital and provide labor.

The benefits and costs of these different types of farms hinge on the tradeoff between efficient work incentives and gains from specialization. The benefit of the family farm organization is that the farmer does not cheat himself. As the sole owner, the incentives for efficient work are perfect. The primary cost of the family farm, however, is a lack of specialization given that the farmer must engage in numerous tasks during each stage of production, and move from one stage to another throughout the year. Such a "generalist" will inevitably be less productive than an equivalent worker in a more specialized firm. Similarly, a family farmer tends to face higher per unit costs of capital because of the limited wealth of a family farm and because a family farm will use capital less intensively.

The partnership farm offers an alternative organization. The benefits of a partnership come in the form of greater productivity because of farmer specialization and lower capital costs. At the same time, adding a partner generates additional costs, in terms of decreased farmer effort because of the imperfect incentives arising from shared ownership.

In a large factory-style corporate farm, the farm's owners share revenues as well as capital and labor costs, but they typically do not provide labor themselves. Instead, specialized wage employees provide labor. The benefits of a large factory-style corporate firm are the increased productivity of a highly specialized labor force and the lower costs of capital. The costs of this regime are the increased incentives for the hired workers to shirk because they are not owners of the farm.

Nature's Impact on Incentives and Specialization

All firms are governed by the tradeoff between work incentives and gains from specialization. This is true of farming as well, but the unique large impact of nature biases the outcome in favor of small firms. One key feature of agriculture is that it involves a living, growing product, which goes through several distinct stages of production. These stages—planting, cultivation, harvesting, and processing for plant crops; breeding, husbandry, feeding, and slaughter for livestock—are largely governed by nature. In principle, there is no reason why a separate farmer could not own each stage.

The decision to keep these various stages of production within the same farm rather than offering them for sale in the market depends on the tradeoff between the gains from specialized stage production and the costs of engaging in market transactions. As with all production, an inter-stage incentive problem emerges because of timing difficulties between stages of production. This timing problem is particularly severe with farming because inventories of the intermediate goods cannot be held given the living nature of the product.

In many cases, small deviations in timing of a task can reduce crop output by relatively large amounts. For example, failure to apply pesticides or to harvest at the right time can be disastrous. Timing causes incentive problems because deviations from optimal time reduce output and because there is uncertainty about when the optimal time will occur. This makes it costly to contract across stages, and an increase in this uncertainty decreases the probability of firm-to-firm contracting between stages.

The problems associated with the timing of stage tasks are not the only incentive effects arising from seasonal forces. Factors such as the number of crop cycles, the length of production stages, and the number of tasks within a stage also influence incentives. When cycles are few, stages are short, random shocks are large, and tasks are also few, there is little to gain from organizing specialized farm labor in a complex, capital-intensive firm. These conditions not only limit the gains from specialization but also make wage labor especially costly to monitor because there is little routine and too few workers for comparison. Both of these forces make family farms more

Figure 1.

The Extent of the Farm in American History

Typical Farm Sequence in 1800

| Stages of Production (Farm Activities) | Genetics, Seedstock | Equipment, Inputs | Planning | Preparing Site | Planting, Breeding | Husbandry, Maintenance | Harvest, Slaughter | Storage | Processing | Marketing, Retailing |

Typical Farm Sequence in 2000

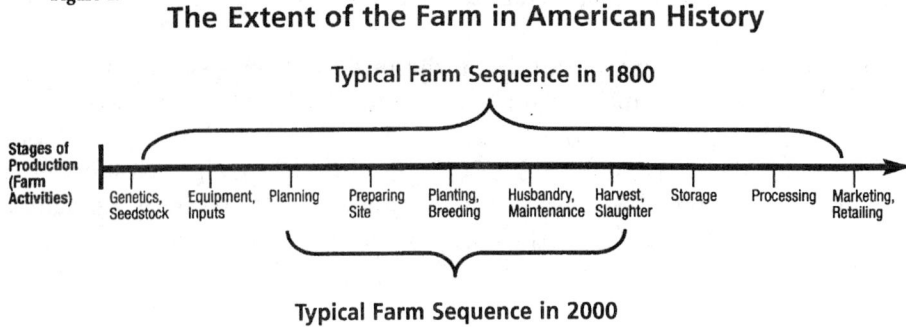

valuable than alternative types of farm organization. In those cases where production is characterized by many cycles, long stages with many tasks, and small shocks, there are gains from specialization and intensive capital that can make large factory-corporate farms the most efficient organization. Where farmers can mitigate seasonality and random shocks to output, farm organizations gravitate toward factory processes and develop the large-scale corporate forms found elsewhere in the economy.

Lesson #1: Family farms are ancient and efficient. The family unit has been the dominant organization in farming since the earliest days of agriculture. Family farms were present in ancient Egypt, Israel, and Mesopotamia and among pre-Columbian American Indians. Owner-cultivated farms have also dominated in Asia, Europe, and Latin America as well as North America. Even in Africa, where land is often owned in common by tribes, farmland is customarily allotted to individual families.

In our framework, family farms are the most efficient organization for crops with many short stages, few tasks, and high susceptibility to unpredictable natural phenomena. Small grain production fits this scenario. Family farms still dominate the production of small grain crops. In 1997, for instance, 80 percent of all wheat farms (nearly a quarter of a million) were sole proprietorship family farms, and nearly all the rest were family partnerships or family corporations. Only 0.6 percent of all wheat sales were derived from non-family corporate farms.

Even the historical variation in farm organization fits this framework. In the United States, the family farm has been less common in southern agriculture than in the north. Plantation agriculture thrived because plantations used one-crop systems that required large amounts of labor on relatively small plots. Compared to grain crops, the plantation crops had a small number of long stages allowing for great gains in specialization and low cost monitoring of labor. American history is riddled with the failed factory farming experiments of ambitious entrepreneurs. From Iowa to California, and from the Dakotas to Texas, early capitalists invested huge sums trying to replicate factory production in agriculture. The great 19th Century "Bonanza" wheat farms—some exceeding 50,000 acres—of the Red River Valley dividing Minnesota and North Dakota were perhaps the greatest of these experiments. The Bonanzas were run by professional managers and used a large, specialized wage labor force to keep the entire production sequence from sod busting to milling flour within the firm. Even though the bonanza farms were hailed as the future of agriculture, nearly all were gone after one generation—most bought out by family farmers.

Coincident with the modern survival of family farming in the midst of rapid industrialization is the dramatic failure of collective farms in centrally planned economies. The widespread famines in China and the Soviet Union are the most notable of the routinely catastrophic outcomes. These failures are not surprising given the limited gains from specialization available in temperate grain agriculture.

Lesson #2: The extent of the farm firm has tended to shrink. Until the late 19th century, the family farm included virtually all stages of farm production, from "farm-making" (clearing land and raising buildings) to processing goods for retail consumption. The family had almost no contact with the market except when the farmer sold (or bartered) products directly to consumers. The main exception was selling grain to gristmills where grains could be easily stored and the mill could operate continuously.

After the early 1800s dramatic changes in technology led to the rise of separate firms that operated throughout the year and specialized in single stages of production. New technologies, such as refrigeration, limited natural forces and allowed seasonal tasks to be performed throughout the year. Overwhelmingly, the new firms engaged in production at either the beginning (handling inputs) or the end (handling products) of the agricultural production sequence. Accordingly, the family farm abandoned these stages and retained control of only the biological growth stages of production. Figure 1 shows how the extent of a family farm has diminished over time.

Even though the family farm has relinquished some stages of production to specialized firms, family farms are sometimes linked to these large firms through vertical integration or long-term contracts. Such connections are particularly strong if timing is important. Vertical coordination occurs when farmers grow crops where precise timing is an important impediment to market transactions. Sugarcane is a good example of a crop with severe timing problems between harvest and processing stages. Because the sugar begins to deteriorate immediately after harvest, decentralized market connections between stages are extremely costly. Thus, virtually all sugarcane production is governed by vertical contracts or by vertical integration with sugar processors; in fact, the grower's harvesting schedule is usually determined by the processor. Such vertical arrangements are rare for highly storable products such as grains.

Lesson #3. The modern livestock industry is different. Livestock production provides the major exception to the dominance of family farming. This has been especially true for broilers, feedlot cattle, and hogs. From 1969 to 1997, there was rising concentration in all livestock industries except cow-calf operations. The general trend has been to remove stock from an open environment and rear them in confinement, often in climate-controlled barns. New technologies—in confinement facilities, disease control, handling, nutrition, and transportation—have reduced seasonal forces by increasing the number of cycles per year and reduced the effects of random shocks from nature. Compared to field crops, livestock production allows for greater control over natural forces because stocks are mobile during growing stages and can often be reared indoors. This control of nature favors factory-style corporate livestock farming.

Feedlot cattle provide a striking example of factory-corporate livestock production, During the last 40 years, the cattle feeding industry has been almost completely transformed into one dominated by large corporate firms that employ highly specialized wage labor. As of 1997 huge firms dominated the industry. Just 230 farms with an average inventory of 30,982 head accounted for over 50 percent of all fed cattle sold, and more than one-half of all cattle sold and receipts came from feedlots organized as corporations. The cow-calf industry, which supplies young feeder cattle to commercial feedlots, is very different. Firms in this industry average only 48 head and are dominated by small, family organizations. The industry is strikingly dispersed with less than one-half of the one percent of the farms having more than 500 head. A cow-calf operation is highly subject to the seasonal forces of nature. Compared to the routine, factory processes in feedlot operations, a cow-calf operation consists of relatively unpredictable short stages (such as calving) that occur only once a year and require on-the-spot decision-making.

The industrialization of poultry production began in the 1930s. Today large, factory-corporate firms produce nearly all broilers. The introduction of antibiotics and other drugs have allowed poultry to be bred, hatched, and grown in highly controlled indoor assembly line environments. At the various stages of production, broiler companies may and often do employ wage laborers who undertake specialized but routine tasks, such as cleaning, feeding, and immunizing. However, the critical "grow out" period of a chicken's life, even using modern technology, is still subject to highly random forces of disease and weather. Large firms routinely contract out growing services to small, family-based "growers." Growers feed and care for the birds until they become large enough for processing. Once chicks have matured, they are returned to the company for processing in large assembly-line facilities that employ hundreds of workers.

Implications for the future. Should we worry about the end of family farming? Will family farms be with us in the 21st century? No and yes. Although the organization of the industry has generally followed a transition from family firms to large, factory-style corporations, most of farming remains a family production activity. Production stages in farming tend to be short, infrequent and require few distinct tasks. This limits the benefits of specialization and makes wage labor especially costly to monitor.

Farm organization will gravitate toward factory processes only when farmers can control the seasonal and random shock effects of nature. When this occurs, farms may develop into the large-scale corporate businesses found elsewhere in the economy. Only if wheat could be grown indoors would wheat farming begin to look like greenhouse farming. Massive factory production in grains, however, seems unlikely.

To be sure, farms will continue to get larger in acreage and output and there will be fewer farm families. But this does not imply a fundamental change in farm structure. This is not a bad thing, just evidence of the invisible hand at work in the organization of an industry.

For More Information

Allen, Douglas W. and Dean Lueck. "The Nature of the Farm." *Journal of Law and Economics 41* (1998): 343–386.

Becker, Gary S. and Kevin M. Murphy, "The Division of Labor, Coordination Costs, and Knowledge." *Quarterly Journal of Economics 107* (1992): 1137–1160.

Brewster, John M. "The Machine Process in Agriculture and Industry." *Journal of Farm Economics* 32 (1950): 69–81.

Coase, Ronald H. "The Nature of the Firm." *Economica* 4 (November 1937): 386–405.

Ellickson, John C. and John M. Brewster. "Technological Advance and the Structure of American Agriculture." *Journal of Farm Economics* 29 (1947): 827–847.

Nerlove, Marc I. "Some Reflections on the Economic Organization of Agriculture: Traditional, Modern, and Transitional." David Martimort ed., pp. 9–30. *Agricultural Markets, Mechanism, Failures and Regulations* Amsterdam: New Holland, 1995.

THE CHANGING NATURE OF EMPLOYEE MOTIVATION

Mike Jackson

With tougher competition and tighter margins, your dealership must perform at its best. Managers are highly concerned with how sales and expenses influence the bottom line. Both of these elements are certainly critical to financial profitability. However, influencing your employees' performance is the most important element of success—and one that is often forgotten in our hectic operating environment.

Motivating employees to top performance is a challenges for all managers. However, you can be more successful in your motivational efforts by:

❏ accepting that people are our most enabling, and limiting, resource;
❏ setting employee goals that are consistent with company goals;
❏ understanding the influence of rewards on employee motivation; and
❏ creating employee rewards that address individual needs.

People: Your Most Valuable Resource

Managers often focus on managing things rather than leading their employees. However, there is a direct link between the financial success of a dealership and the people it employs. Therefore, your success as a retailer lies in the hands of your employees. Are they ready to implement changes in our strategy? How well can your employees sell products and services? How do they add value to your customers?

The quality of your products and service and the value you provide to customers are only as good as the quality of your employees. If your employees are unable to sell solutions and build solid relationships with customers, your customers will take their business elsewhere. Successful retailers accept that their profitability is directly linked to their employee's performance and channel their management time accordingly. While other daily activities may seem more urgent, make time for leading and motivating. But how should you use that time? What should you do?

Setting and Aligning Goals

One of the most essential elements of motivating employees to top performance is setting clear goals. As a manager, you understand the value of individual goals for your employees. But truly effective goals lend not only to improvements in personal performance but to improvements in your dealership capabilities as a whole.

So, before setting individual goals with your employees, it is essential that you communicate your dealership's vision, goals and priorities Make certain that individuals understand how your dealership's goals will impact their jobs and performance and then work with employees to set goals that are complementary with corporate goals. To create effective goals, make certain the goals:

❑ will achieve strategic results:
❑ target improvements in individual performance; and
❑ are measurable and realistic.

This step is essential because it allows employees to "fit" their priorities to company priorities. People can't get motivated to work for company's success unless they understand their role in that success. That is particularly true in times of dramatic change.

Motivating Employees to Top Performance

Once goals are in place, how can you keep employees working toward top performance? Motivating employees is a challenge that all managers face. Is it possible to "motivate" the people you supervise? Not really—although you can set up a system of rewards and penalties for performance. However, inner motivation, that drive that causes some people to do a great job, comes from within. Motivation isn't something you do to people, but something they do for themselves.

So as a manager, how can you influence motivation? First, it is essential to understand that each of your employees is motivated by his or her needs. Therefore, as a manager, you can influence motivation by supplying rewards that address employee needs. It is important to remember that there are two main types of motivation:

❑ extensive motivation and
❑ intrinsic motivation.

As a manager, you can influence extensive motivation through salary, benefits, and bonuses and other types of compensation. However, intrinsic motivation—feelings of personal satisfaction and value—are generated by individuals. As a manager, you can influence intrinsic motivation through coaching and other activities that help employees realize their value to your dealership.

Working with Intrinsic Motivation

Many managers mistakenly believe that employees are only motivated by extrinsic rewards. However, many employees take great satisfaction in their jobs and are highly motivated by more intangible rewards, such as feedback about a job well-done.

As a manager, your challenge is to first, understand what motivates your employees, and then to create a reward package that addresses their needs. Work with employees to generate a system of rewards that addresses their individual needs—not a one-size-fits-all benefits scheme.

Managing in the 21st Century

Just as each of your employees has a different work style, each of your employees is motivated by different factors. The key to employee motivation is helping your employees set realistic performance goals and creating reward systems that incite them to top performance. By providing your employees with a comprehensive reward system that addresses their individual needs, you ensure that you are managing all aspects of your bottom line.

CHALLENGES FOR THE NEW MILLENNIUM

Larry Daniel

Heading into the 1990s, the business geographics landscape was relatively barren. Mapping technology hadn't reached the PC, and data was relatively sparse and expensive.

For sure, there were some organizations dabbling with DIME files and the 1980 Census; there were even a few that prospered by specializing with this processing at a time when others hadn't the wherewithal. But the general state of information processing required such intense resources that innovative or entrepreneurial efforts in the area were few.

How times have changed. In the span of a decade, geodemographic analysis is now within the reach of all. The 1990 Census and TIGER efforts were pivotal to setting the stage for growth, and the developments with faster and less expensive computer hardware opened the playing field to a much broader range of dabblers, innovators and champions.

Now as we enter 2000, the industry is joined by Microsoft Corp., which recently launched a desktop geographic information system (GIS) and business tool for $109. Other companies, such as MapInfo Corp., Claritas Inc. and SRC, are now offering pay-as-you-go demographics for even less on the Web.

The state of the industry per 1989 is certainly no longer. Software is easy to use, services are abundant, datasets are exploding, and prices are falling to commodity levels.

Are all the revolutionary and evolutionary changes complete? If we accept time as our teacher, we know that it can't be so. Technology firms and solutions are in a constant state of flux and must constantly reinvent themselves to survive. If anything, the pressures for growth and change are only intensifying. More and more technologies are competing for the corporate dollar, and the competition between geographic and other types of processing will likely spawn more innovation.

The challenges to the industry's growth are abundant and include, but aren't necessarily dominated solely by, whether we all can hitch onto the Internet economy. Below, we classify some of the challenges related to business thought, platform issues and applications as we visit some of the battlegrounds more closely.

Business Thought

In the oft-cited "Crossing the Chasm," Geoffrey Moore makes the case that industries face something of a challenge in pushing technology products beyond a stage of being interesting only for innovators and into a stage of interests for a larger population of common adopters. Within the business geographics industry, it isn't unusual to hear or become engaged in discussions about whether geotechnologies have indeed "crossed the chasm."

There is, as one might imagine, no clear consensus. The comments in these discussions vary quickly depending on what aspect of the industry the participants come from. In its article, "It's a Map, Map, Map World" [Oct. 4, 1999, page 25], the *Industry Standard* stated that online mapping services "have become de rigeur for self-respecting portal sites."

The *Standard* cited a statistic that MapQuest's service creates 150 million maps a month in comparison to 100 million paper maps that are sold each year. Could it be that mapping has indeed arrived?

A saltier observation is that mapping has arrived chiefly where it can serve audiences with economies of scale. Are 1,000 maps a day considered a suitable return for firms that invest in GIS in order to provide Net surfers seeking the closest store to their homes? Probably.

Is $1,000 a day accepted as a reasonable investment into consulting services that produce site selection models for the company to determine where the firm should place its stores? Many might say yes, but many more are uncertain. Geographic information, apart from that relating to how to get somewhere, still seems greatly undervalued.

The exciting observation, though, is that business geographics can now leverage these opportunities. The potential economies created by the Internet and other pervasive computing platforms as a distribution media now provide an economic route to introduce geography. If one has ever tried selling site selection or territory management services on their own merits, you learn that specialized geographic applications such as these hinge on proving the cost effectiveness of the project.

Now there is potentially a whole new spin on investments into geographic tools and data. It can be about empowering one's clients to use geographic intelligence in both a very widespread and very deep manner.

This repositioning is exactly what venture capital investors seem to be particularly attentive to. As stated in *Computerworld* [Nov. 15, 1999], "Following the money leads [one to observe] to the major trend within ecommerce: a focus on using Internet technology to completely revamp how companies share and use information —[attaining] information liquidity."

The objective is increasingly to create data flows between information technologies to create content, an objective for which desktop mapping is very well suited to contribute.

New Platforms: the Internet

Is it really possible that most of us were unconcerned about the Internet tide five years ago? In less time than it takes some users to upgrade to a new version of application software, a whole industry has taken technology and our economy by storm.

Ponder the numbers below to get a sense of how it has rocked the business and consumer worlds:

- ❏ The 1998 Internet economy achieved $301 billion in revenues.
- ❏ In 1998, e-commerce spending was about $50 billion. That number is expected to jump to $1.3 trillion in 2003.
- ❏ Yahoo had 32.3 million visitors in July 1999.
- ❏ The Web now has more than 800 million separate pages. The largest search engine covers only 16% of these pages.
- ❏ Internet mergers and acquisitions through the first half of 1999 totaled $43.4 billion.
- ❏ By 2003, 14 million Americans will likely access the Internet via TV.

At MapInfo's MapWorld Conference in Miami this past summer, the featured speaker on a track called Technology Trends announced, "Well, there's really only one thing to talk about: the Internet." Was he wrong? He may have overstated the issue a bit, but not much.

The business geographics challenge with the monster opportunity involves thinking clearly about how we unhitch ourselves from our PC/workstation past and link ourselves to the cyber highway. The geographics community grew up with specialized systems, powered to provide the graphics that were essential to visualizing our products. But the links have changed and the bandwidth for streaming GIS/mapping elsewhere is expanding.

Lucent Technologies reported research, for example, that indicated new forms of fiber optics would soon deliver the entire Internet's weekly transmission needs within seconds. Capacity, once thought to loom ahead as a barrier for application, seems to be less and less an issue. The infrastructure improvements enable us to ignore specialized hardware and fit into the communication media that is rapidly changing our world.

A key challenge is for mapping types to identify and accommodate the burgeoning need for content on portals. When we can find 800 million pages on the Web, the general need on the Internet is for sites that can focus the attention of surfers and link users with common interests.

A recent *Computerworld* survey (reported Aug. 9, 1999) found that 72% of managers from mid-sized to large companies were involved with portal development. Portals deliver company-specific information to provide access to human resources applications, for example, and to provide a focal point for employees to launch into needs for other information.

A major growth path would appear to involve determining how our integrating and visual technology can become a greater force for those portals under construction, e.g., not just cool Web sites, but integral pieces of different industry and company portals.

Projections for business-to-business e-commerce also warrant attention. Forrester Research reported in the Sept. 13, 1999, edition of *The Industry Standard* that online revenues for motor vehicles, for example, will grow from $9.3 billion in 1999 to $212.9 billion in 2003. Demand for utilities will grow from $15.4 billion to $169.5 billion, and for computer and electronics, the market will increase from $50.4 billion to $395.3 billion.

One wonders what type of opportunities await the business geographics organization that can determine how to add value in these types of online transactions.

There are tremendous opportunities to seize the moment and provide geographic applications for the Web. The challenge is to do so without overlooking the opportunities for introducing entirely new types of geographic content and depth. As an industry, we can all rush to produce locator maps, or we can move into a more prominent role with information management.

The area with the most intuitive content-oriented opportunity may be related to transporting value-added information—GIS' core specialty—via the Internet, corporate intranets or host-driven extranets. There are intriguing opportunities for GIS firms to play leading roles in data mining, customer relationship management and knowledge management activities.

In a June 21, 1999, supplement to *Computerworld,* International Data Corp., a Framingham, Mass., technology market research company, demonstrated the value of pulling disconnected communities, such as the service organization, corporate offices, sales, and research and development, together. What better tools exist for accomplishing this than deploying GIS as a tool to manage common data via the Web?

Of course, there are a few intriguing developments besides the Web. The Internet may be the biggest, but it isn't the only name in the changing platform game.

The biggest force at the Fall '99 Complex show, for example, was wireless communications. Other noteworthy growth opportunities will emerge with in-vehicle navigation and the new generation of mobile devices.

An interesting notion related to business geographics is the emerging integration between wireless communications and handheld computing. Other mobile information appliances, such as Palmtops, data viewers and smart phones, conjure up thoughts of James Bond-like possibilities.

The demand prospects for these products is staggering. Imagine remote retrieval of real estate data or customer information in maps, charts and geographic analysis as needed, on demand by whoever on your team is most in need of direction.

A key challenge with respect to these platforms is to recognize them foremost as vehicles for greater information distribution: Can we demonstrate to the masses that we offer more than just maps? If we integrate additional components of technology into these exciting platforms, we can make a profound business impact by empowering end users to perform their roles not only with georeferencing but also with greater geographic knowledge.

Applications: 3-D Multimedia

If there is one idea on which those within the industry seem to converge, it is that we are on a collision course with more exciting, more dramatic forms of 3-D multimedia GIS. In the January 1995 issue of *Business Geographics,* David Wamsley, who was then senior analyst for Hilton Gaming Corp., Las Vegas, speculated that the future of business geographics would be like a video game.

"I will be wearing my virtual-reality glasses with sensor gloves and I will be racing through key markets on my screen, shooting my way through decisions that related to how I might over-take my enemy on the horizon," Wamsley envisioned. Interestingly enough, some of those visions no longer seem far fetched.

It may be coincidence, but several GIS journals highlighted the multimedia 3-D developments this past fall. One was *ArcUser,* published by ESRI Inc., which featured the article "Three Dimensional Databases in ArcView GIS" in its October–December 1999 edition. Author Daniel Elroi stated, "Chances are that a common experience for most GIS professionals is having to explain that GIS is two-and-a-half dimensional. No, the earth isn't flat; nor should an information system that describes and analyzes it be flat. "

"The Virtual Reality Race" by Peter Warner in the November/December 1999 issue of *Imaging Notes* stated: "Price/performance breakthroughs in GIS software and desktop computers, along with the promise of high-resolution [satellite] imagery, will help 3-D modeling and simulation applications expand into scores of new applications, including real estate, legal, insurance, telecommunications, wireless, business presentation, travel, tourism, recreation and media."

The reality is that software is maturing, computational power is increasing, and related datasets are becoming available to make the potential for very sophisticated cyber methods of simulating a city. Longtimers in the industry might remember how Jerry Robinson began a GeoVan venture in the mid-1990s to videotape and georeference U.S. communities for display in various types of GIS applications.

Our research can't locate or determine how that venture ended up, but the reality might be that it is time for it to cycle about in a newer format. Imagine yourself as a vice president of real estate able to visualize all prospective site expansions without having to either travel to the sites or glance through photographs. One might "drive through" the market at helicopter level to view the layout of the prospective locations, evaluate nearby demographics and examine the condition of the competition.

The application is visionary and certainly challenging. For the moment, no application of this type has reached the mass business market. Once it receives the capitalization, development and positioning, the application would seem on course to change the way that users think of GIS thereafter. Once 3-D becomes vogue, could we ever go back to 2-D maps again?

Other Growth Opportunities

Perhaps the near-term challenge in the 3-D multimedia direction is to take a step toward perfecting the geographic multimedia presentation itself (without 3-D). As William Cartwright discussed in "Multimedia Creates New Map Products" *[GeoWorld,* October 1999], "The mapmaker's palette now contain ways to depict and deliver geospatial information in more timely, resourceful and exciting ways."

By using dynamic HTML, sound and mapping, animation, charts and more complete descriptive text, it would appear that business geographics presentations could go far beyond their all-too-frequent state of just being maps.

Another opportunity, although certainly more mundane, relates to the simple explosion of new datasets. If the coming decade emphasizes content, we need to eliminate the gaps where our lack of data prevents those using business geographics from presenting an accurate picture of the areas we intend to analyze.

Data has all too frequently stymied the vision of grandiose GIS projects. New and more comprehensive sources of business data, more advanced consumer behavior panels and continued advances with population estimation techniques, both abroad and in the U.S., seem key to establishing and/or maintaining business geographics' position as a leading-edge technology. As an industry, we are challenged to ensure that our databases represent reality.

We will be forced to address many challenges in trying to extend the influence of business geographics technology. The greatest challenge, though, may simply be crossing our chasm without isolating ourselves all over again.

Will we simply enable companies to visualize geographic objects in centuries-old formats or empower end users to see data as they never did before through GIS-generated content? We have a unique integrating edge. It's time to demonstrate how and why it is worthwhile to deploy it on a broader basis.

PLANT AUTOMATION MIGHT BE CLOSER THAN YOU THINK

Bill Tindall

Purchasing a fertilizer and crop protection handling system that interfaces with agronomy and accounting software, while streamlining your ag retail operation with sensors, meters, and computer hardware and software, might be closer to reality than you think.

John Christian, a partner in Green Valley Ag in Caledonia, MI, says, "We wanted to find an easy way to handle all of our liquids from one location, and when we sat down with a sales rep from Murray Equipment Inc. (Fort Wayne, IN), we were surprised at the positive economics of investing in an automated handling system. Besides the advantage of having one man do what it took two to do before, we anticipate a fairly quick payoff on this new system."

With a mix board that "isn't really fancy," he says, but gets the job done, Green Valley is clearly able to better serve its grower customers with the addition of the automated system. Nearly 40 percent of the sales of this retail operation, which services an approximate 30-mile radius just south of Grand Rapids, are to growers who apply with their own equipment.

"Quick load-out lets our growers get into the fields with fertilizer and ag chemicals that are going to make their farms profitable," explains Christian.

Additionally, Green Valley custom-applies with three spray rigs, two Silver Wheels units and a Wilmar Eagle.

In Clunette, IN, Manager Dave Truex of Clunette Elevator says, "In our mix plant, we run three vats, each controlled by a separate unit for different analyses." After the information is entered in, the system figures percentages to meet each grower's specific needs. This is particularly important in meeting the needs of GPS customers," he adds.

Air-controlled valves keep the system moving and offer plenty of flexibility. With the Rice Lakes IQ 800 Controller used by Murray Equipment, the order goes to the plant, and recommended products are keyed into the computer. "Once we do that," explains Truex, "we let the machine loose."

Products are brought in by weight, go through one at a time until the blend is made, and wait in the pause mode until the product is ready for use. Another handy feature is that starter inventory can be sent back to storage until it is needed.

Clunette Elevator also freed one man's time by automating its facility. "Because we don't have to stand on the load-out area and manually operate nozzles and valves, we now have an opportunity to talk to customers and generate additional business," Truex says.

"Many times a simple update will lead to further automation," he adds. "It has taken us six years to get fully automated. First we had the Murray plant installed. We liked how it worked, so we went back and updated and automated the old plant. I think it makes good strategic sense to leave room for expansion in this business."

Look Around for the Right System

Brad Wilson and Doug Deno with Wilson Fertilizer in Brook, IN, looked around and visited other retail sites before making the choice to automate their "weed and feed" business with Murray's Touch Screen. "We liked what we saw with the Touch Screen but didn't think it would be an option," explains Wilson. "However, when they ran the figures, we realized it was affordable."

Before, Wilson Fertilizer pumped 28 percent into a mix tank, added chemicals and agitated the mix. Then an employee pumped it again into nurse tanks or application floaters. Now, at the rate of 300 gallons per minute, the company's Murray automated system will load a 3,000-gallon unit in 10 minutes. "We estimate that our time savings alone is 10 minutes per load," Wilson says.

In addition to the time savings, the retail site saves on manpower. Where it took two men on the loadout pad previously, one is all it takes now. "Additionally, that one man can do two or three other things while the automated system loads," he says.

To accommodate the increased product generated by the automated system, Wilson bought an additional applicator this year. "We have plans to purchase another tender to keep up with the increased volume," Wilson says.

Meets Demands of Variables

Jerry Beuning of UAP Midwest in Sauk Centre, MN, says that peace of mind is foremost in the list of benefits derived from the system the business had installed three years ago by Kahler Automation. "The accuracy is just unbeatable," he explains.

Temperatures vary and products come in an array of viscosities. "Whether we are adding atrazine, which tends to be thicker, or Harness, which tends to be thinner, this system makes the necessary adjustments," says Beuning.

The automation gives the retail site the diversity and flexibility to serve its customer base ranging from 200-acre dairy farmers to 4,000-acre cash grain and livestock farmers.

The Kahler system completely and automatically weighs and dispenses each load. "We simply hit 'mix' on the computer, designate the gallons of each product desired, and push 'go'," explains Beuning. "It weighs and injects the mix into the fill line to load either nurse equipment or applicators." The retailer also conveniently fills mini bulk 100- and 200-gallon shuttle tanks with the fully automated system. "One real asset," explains Beuning, "is that at least half of our drivers can now load themselves. They just hit the 'Go Load' button and the automation does the rest."

Accuracy is also an important consideration for UAP Midwest/Sauk Centre, a site that sells 300,000 gallons of material each spring. "With 15 machines and tender trucks going through a load pad all day, every day, we have to be accurate," says Beuning. "Accounting alone is a big task."

Wholesale business is an important aspect of UAP Midwest. It provides product to all of Minnesota north of the Twin Cities, excluding the Red River Valley. In addition, it does custom application for other retailers, which adds to its concern over accuracy at the load-out pad.

Agronomy Manager Charlie Head says that Kahler's automated system is a good PR tool that benefits both the business and the entire industry.

"We love to give tours of our facility to show the technology behind our services," he concludes.

Automation Systems Offer Versatility to Retailers of All Sizes

Like Clunette Elevator, Cropmate Company in Cherokee, IA, has plans to further automate its system over the next two to three years. "We like the Kahler Mass Flow unit," says owner Mark Braunschweig.

"It saves both time and labor," explains Braunschweig. "Not only can we break away from the loadout pad to take phone calls and do other things, but the automated air valves are capable of cleaning the lines in 20 seconds. We hardly see a drip anymore."

Future plans call for adding automation equipment to include chemicals.

"We also plan to automate our invoicing," adds Braunschweig.

Automating at Agriliance Agronomy Center in Cozad, NE, where they apply product to 100,000 acres each year, has more than doubled its load-out speed. Its decision to automate with FarmChem's system was driven by inefficiencies in both labor and speed of blending.

"We completed installation in March last year, so we have a year of experience under our belts," says Agronomy Manager Mike Bellamy. "For the first time, we didn't have trucks sitting on the lot, waiting to be filled."

Previously, the site used two tub mixers. "We pumped product in, mixed it, and pumped it back out again," says Bellamy. "Now, we can walk away once information is programmed in." Automated with two mass flow units, speed has more than doubled.

The system operates from a Junge control panel. Three boxes, one to operate the mix tub, one to run the chemical system and one to control the mass flow, meet the needs of 10 flotation rigs, four row-crop applicators and six three-wheelers. Three one-million-gallon tanks, all controlled by the computerized system, are utilized for extended storage.

They are located approximately, 1,400 feet from the load pad but can be easily accessed by the computer system when needed. "This system gives us an opportunity to prepare product in advance and take advantage of price appreciation," says Bellamy.

"We still have some manual entries," notes Bellamy. However, Agriliance Agronomy has the capability to add automation as it goes. "Our plan is to grow into additional automation," he concludes.

With $50 million in sales, nearly half of that derived from fertilizer and crop protection sales, Central Farmer's Co-op, headquartered in O'Neill, NE, has to handle large volumes of product in a short period of time.

It covers more than 600,000 acres in a geographic span that reaches 200 miles east to west and 150 miles north to south.

The company built its first automated facility in 1994, a state-of-the-art site utilizing FarmChem plant automation equipment. About a year ago, it renovated its latest facility, in Tilden, NE, with Farm-Chem's automated system. This unit can load semi trailers in six minutes.

"We have always felt strongly about meeting environmental standards, Karl Hensley, agronomy division manager at Central Farmer's Co-op, says.

The other advantage of the automation system is the ability to enter data, such as products desired, acres to be serviced and customer information, prior to the hectic application season. "Everything works from the flip of a switch," Hensley says. "Not only are we able to operate with greater efficiency, but we can better meet our customer's needs by freeing professional time and allowing the automation equipment to perform the mechanical functions."

Central Farmer's Co-op doesn't regret making investments in automating its facilities. "We toured several sites before installing our own," explains Hensley. "The ideas we got from other retailers allowed us to tailor the equipment to fit our specific needs. If there is one recommendation I would offer, in looking back at our experience in updating plants, it would be to take the time to see other people's facilities."

Scouting Out Profits

K. Elliott Nowels

When Alan O'Neal was concerned about farmers learning where his new UAP (United Agri-Products) Carolinas operation was located after Hurricane Fran destroyed the old outlet, he just rang the bell. The dinner bell.

Every Wednesday in the winter time, O'Neal put the food on the table, and farmers found the facility.

"We try not only to feed them, but we try to have some good meetings, too," says O'Neal of the effort that continues even though it's been two years since the facility's relocation. "To help them, that's what we're after. We want to help them produce a better crop."

The hurricane blew the business down in Sept. 1996. They didn't have a new facility until the following September. How did they handle that growing season in between?

"Luckily, my salesman, Scott Sinclair and his parents have a produce pack house, and they took us in as a tenant," remembers O'Neal. "We had to do several things to get the warehouse safe and had to develop a new plan of action. Then we pulled a mobile trailer in and made it work."

Weathering the storms, whether natural or political, is a fact of life for O'Neal and his customers in this part of North Carolina. O'Neal operates the UAP Carolinas outlet in Clinton, N.C., located about 60 miles south of Raleigh and 60 miles inland from the Atlantic Ocean. To the east is Onslow Bay, a place where the coastline curves from south to east, forming a perfect scoop to catch the brunt of hurricanes hurrying northwest from the Caribbean. When the weather's right, it's a location fit for agricultural diversity. O'Neal and his UAP team serve some 300 farmer-customers and almost as many cropping patterns.

"Blueberries, peppers, cucumbers, squash, greens, all of the greens—the kale, the kohlrabi, mustard, turnips, the whole deal," lists O'Neal, taking a breath to continue. "A lot of sweet potatoes, sweet corn, watermelons, cantaloupe, tomatoes, pickles, long green cucumbers. Did I mention cucumbers? And bell peppers—then there's some jalapeno and hot pepper also." All this and then the mainstay: tobacco.

"Tobacco is No. 1, and cotton would be No. 2," says O'Neal, "then vegetables probably No. 3 and then corn and soybeans."

Tobacco, says O'Neal is like the center of a wheel. Everything else in North Carolina agriculture still spins around tobacco. As O'Neill describes his own trade territory, the reason is obvious.

"Sixty thousand acres of cotton and eight or nine thousand acres of tobacco," O'Neal muses. "And there's more money in the tobacco there than there is in the cotton. Net profits."

And with tobacco about as unpopular from a national policy standpoint as a legal plant can get, many farmers aren't sure which way to turn. Gross income on an acre of tobacco will top $3,000. Compared to that, an average crop of 24 bushels per acre soybeans isn't very exciting. And the continued litigation over tobacco upsets O'Neal's UAP customers, who can't see a suitable alternative to a crop that's been grown in this territory for more than 300 years. Not even sweet potatoes?

"Unless you start going to the local steak house and start ordering the sweet potato instead of a baked potato," O'Neal smiles, "then we could do something about that. But until that happens, sweet potatoes—the market's mature there. So I don't know if there is really any viable option."

O'Neal, like others in North Carolina, laments that the state is simply ill-equipped to grow enough grain to feed the livestock population it boasts. By some accounts, if North Carolina was a country, it would be the fourth largest importer of feed grains in the world.

"If every acre that's in production now was corn, we could grow enough corn to run the (livestock operations) about three months." Dealing with the winds of economic change here may be more difficult for O'Neal and his growers than dealing with the gales off the Atlantic. In the eye of these storms, O'Neal ponders where to put his team's focus.

"I have trouble seeing that custom application would be profitable for us in our situation here," says O'Neal, who believes in doing a few things, owning his niche and clearly focusing his people.

"Instead of trying to be everything to everybody, we try to be better at some things than others," says O'Neal. "We see our strong suit as doing a lot of field work consulting-type activities and crop protection."

O'Neal is centering his efforts on the knowledge business by serving growers and their advisers with the best information they can get anywhere. An information service called mPower[3], which UAP is marketing nationally, is offering weather information and data management tools geared to running field specific crop models. UAP Carolinas is working locally to prove its worth.

"I'm working real strong with that right now through two local consultants to help them improve their operations," explains O'Neal. "I'm not sure that the farmers are ready for that product directly. But I see where a consultant could use it."

O'Neal feels that many of the higher-dollar crops in his area can support these knowledge-building databases.

"If you're a cotton consultant, you're talking about a $20 program, full program to consult an acre of cotton," he explains. "You take our mPower³ product and it's somewhere in the area of $9 per acre, and it's an easier way for the consultant to gather and compile his data."

O'Neal then sees consultants carrying the ball through with growers, figuring the $9 and passing some of the value and cost along to the grower at a rate that might be $25 per acre.

"That's instead of us trying to go out and sell the farmer $9 for something that he may not really use a whole lot," says O'Neal. "The farmers want somebody to hash through all the data and tell them what they need to do and when they need to do it."

Here, as elsewhere in North America, agricultural expectations are geared to fewer farmers, larger farms and more "dirt to dinner plate" vertical integration. O'Neal feels that his company, with its parent, ConAgra, is in a good position to profit.

"And some people are going to like it, and some people are not going to like it. The bottom line is some things are going to have to change here," says O'Neal. So the business plan at UAP Carolinas involves sticking close to customers, saving them time and minimizing overhead back at the warehouse.

"We try to concentrate on the seed business No. 1, and then the crop protection products that go with it," says O'Neal, "instead of spending a lot of time running up and down the road with trucks. With the margins the way they are in the business today, more overhead is not really the answer."

O'Neal is betting on his people to carve out success for his business and its customers in coming years. The right people in the right place at the right time.

"That's the key to it," he says. "From the salespeople to the secretary to the truck driver to the warehouse manager. We have good people and I think that's the key for the farmer now."

FOR CALIFORNIA FARMWORKERS, FUTURE HOLDS LITTLE PROSPECT FOR CHANGE

Philip L. Martin and J. Edward Taylor

Agriculture is a major employer in California. Some 800,000 to 900,000 people work for wages at some time during a typical year on California farms. Only about half of those work year-round so that farmworkers represent just 3% of California's average 14 million wage and salary workers. Most farmworkers in California are seasonally employed on one farm for less than 6 months each year, and earn a quarter of the average factory worker's annual salary. The vast majority are Hispanic immigrants. During the next quarter century, these trends are likely to continue, with the farm labor market becoming increasingly isolated from the mainstream. An alternative scenario is that strong unions and government regulations could transform farm work into an occupation that can provide a career and support a family. Immigration policy will play a critical role in determining the characteristics of California farmworkers in the 21st century.

Most major farm labor debates at the dawn of the 21st century involve arguments about the proper role of government in the farm labor market. How should trade and research policies influence farmer decisions on what crops to grow and how to harvest them? How easy should it be for farmers to employ foreigners as guest workers? What labor and immigration laws should apply to the farm labor market, and how active should governments be in enforcing these laws?

Current debates about the farm labor market can be framed by two extremes. One scenario imagines that hired farmworkers will increasingly be recently arrived immigrants, so that the farm labor market will be further isolated from other U.S. labor markets. Under this scenario, the seasonal workers of 2020 are being born today in Mexico and Central America. The other extreme imagines unions or government regulation making farm work an occupation that can provide a career and support a family. The actions of farmers, workers and government will determine where the reality is likely to fall between these extremes.

This paper surveys the farm labor market at the beginning of the 21st century and outlines its likely evolution. The number and characteristics of farmworkers played a major role in shaping 20th-century agriculture and the farm labor market, and the farm labor supply is likely to continue to do so in the 21st century. At the beginning of the 20th century, farmers worried about whether Chinese and Japanese farmworkers would continue to be available; at the end of the 20th century, farmers worry about the future availability of Mexican farmworkers.

During a typical year, the 35,000 farm employers in California, including crop growers, livestock farmers, custom harvesters and farm labor contractors (FLCs), hire 800,000 to 900,000 individuals. Most farm employers are native-born, non-Hispanic whites, while most farmworkers are Hispanic immigrants.

Farmworkers' average hourly earnings are about half of average manufacturing wages, $6 to $8 versus $12 to $14 per hour. The average hourly earnings reported by the U.S. Department of Agriculture in *Farm Labor* publication include the earnings of supervisors, which raises average hourly earning figures (USDA 1999). In 1999, for example, average hourly earnings for all hired workers in California were $7.88. However, average hourly earnings for field workers were lower: $7.18. Farmworkers average about 1,000 hours of work per year, about half as many as manufacturing workers. As a result, farmworkers in California have annual earnings that are one-fourth of the $24,000 to $28,000 average of factory workers.

Farm Labor Market Characteristics

Four characteristics distinguish the farm labor market in California:

❏ The farm labor market is dominated by specialized enterprises with highly seasonal labor demands—peak employment can be 20 to 30 times greater than trough employment.

❏ Since 1960, labor-saving technologies have not reduced the overall demand for low-skill farmworkers. Instead, increased production of labor-intensive crops and the shift of some nonfarm packing work to the fields (for example, field packing) increased the average monthly employment of farmworkers in the 1990s.

❏ Most farmworkers are immigrants, and virtually all new entrants to the farm work force were born abroad. U.S.-born workers have almost entirely disappeared from the farm labor market.

❏ Farmworker earnings are among the lowest of any segment of the U.S. work force, reflecting relatively low wages and less than full-time employment. Relatively few farmworkers receive fringe benefits such as health insurance so the farm-non-farm gap in total compensation (earnings plus fringe benefits) widened in the 1980s and 1990s.

Farming in California is often compared to manufacturing. Most farmworkers in California are employed in open-air enterprises that turn raw materials into finished products. A "farm factory" brings together people, land, water and machines to transform seeds into crops. Because the agricultural production process is biological, farm factories face risks that do not arise in manufacturing production processes governed by engineering relationships.

California agriculture is dominated by specialized enterprises that often hire hundreds of workers for a 3-week harvest. Unlike the typical Midwestern family farmer, who does most of his own farm's work, the managers responsible for California's labor-intensive crops rarely hand-harvest themselves. A familiar adage captures many of the differences between California agriculture and Midwestern family farms: California agriculture is a business, not a way of life.

California fruits and vegetables do not ripen uniformly, so the peak demand for labor shifts around the state in a manner that mirrors harvest activities. Harvest activity occurs year-round, beginning with the winter vegetable harvest in Southern California and the winter citrus harvest in the San Joaquin Valley and ending with late olive and kiwi harvests in October.

In late fall and early winter, some workers migrate to Southern California and Arizona for the winter vegetable harvest, and others return to Mexico, but most remain in the areas where they did farm work, jobless and waiting for a new season to start.

Workers willing to follow the ripening crops can find 8 to 10 months of harvest work each year. However, relatively few workers follow the ripening crops in California. A 1965 survey found that 30% of the workers migrated from one of California's farming regions to another (California Assembly 1969), and a 1981 survey of Tulare County farmworkers found that only 20% had to establish a temporary residence away from their usual home because a farm job took them beyond commuting distance (Mines and Kearney 1982). The National Agricultural Workers Survey, conducted annually, reported that 20% to 40% of California crop workers interviewed would be willing to or had traveled beyond daily commuting distance from their homes to do farm work (USDOL 1998, Gabbard et al. 1994).

The number of farm jobs in California has been remarkably stable since the 1960s, and it rose in the 1990s (Fig. 1). The loss of jobs due to picking a crop by machine rather than by hand in many commodities has been offset by the growth of jobs in other farm commodities and the substitution of hired workers for family workers on many farms. During the 1960s, when the processing-tomato harvest was mechanized, it was widely expected that most crops grown in California would be harvested mechanically by 1975. This did not happen, largely because workers were generally available and because of the costs involved in adapting plants and machines for hand-harvesting some perishable commodities.

Labor in the 1990s

Most California farmworkers are Hispanic immigrants. The National Agricultural Workers Survey interviewed 1,885 crop workers employed in nine California counties between 1995 and 1997, and found that 95% were foreign-born, including 91% who were born in Mexico (fig. 2). About 53% of those interviewed had been in the United States for less than 5 years, and 26% for less than 2 years. About 48% were legal immigrants and 42% were unauthorized (USDOL 1998).

Most farmworkers are young men with families. In 1995 through 1997, about 82% of California crop workers were men. The median age of farmworkers was 30, 31% were under 24, and 63% were under 34. About 61% of crop workers were married, and most married workers had families, with an average of three children each. About 60% of farmworkers in the mid-1990s had their families living with them while they did farm work in California; 40% left their families outside the United States. Two-thirds of the workers interviewed had less than 8 years of education, which they usually acquired abroad. Their median years of schooling is 6 (USDOL 1998).

California farmworkers averaged 23 weeks of farm work a year in the mid-1990s, 3 weeks of nonfarm work, and 26 weeks without farm work. In most cases, time not working is spent outside the United States. Most of the workers interviewed (91%) were employed in fruits and vegetables. Of the jobs performed by sample workers in the previous 12 months, about 70% were pruning, irrigating and other nonharvest operations, and 31% were harvesting. Hours of work averaged 42 a week, and average hourly earnings were $5.69. Most interviewed workers had low incomes; 55% earned less than $7,500 in 1996 (USDOL 1998).

A Century of Farm Work

These characteristics of farmworkers are not new. Farmworkers have generally been newcomers to the state with few nonfarm job options because they lacked the language skills and contacts to move out of the farm labor market. California farm-labor history is the story of waves of newcomers entering the state to do farm work, and then returning to their country of origin or moving into nonfarm jobs. Farm-worker's children who are educated in California generally refuse to follow their parents into the fields, so that most entry farmworkers have been raised outside the state (Martin 1996).

The state's growers have had a keen interest in U.S. immigration policy since labor-intensive fruit and vegetable farming developed in the 1880s. Farmers feared that they would have to slow the planting of trees and vines in the 1880s, after the federal government ended Chinese immigration in 1883. However, labor became available from Japan, and plantings of labor-intensive crops tripled in the 1890s. Worries about unskilled immigrants in cities led the United States in 1917 to exclude immigrants over 16 who could not read in any language. California farmers asked the U.S. government to exempt Mexicans coming to work on farms and railroads, and Mexicans soon dominated the farm work force in many areas.

Mexicans stopped migrating to the United States to do farm work in the 1930s, and many already in California were forced to return to Mexico during the Depression. After 1935, small farmers from the Midwest and South began arriving in California, hoping to begin as hired-hand farmworkers and work their way up the agricultural job ladder to become farmers in their own right. Most did not, and the conditions under which some lived inspired an outpouring of farm-labor literature, including John Steinbeck's *The Grapes of Wrath* in 1940.

Numbers of California farmworkers, 1992–1997. Source: Employment Development Department.

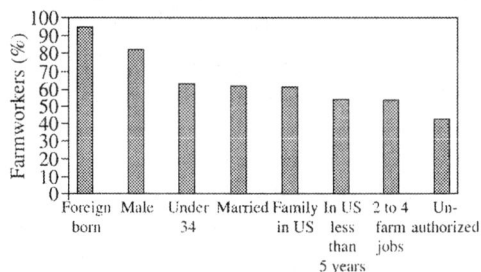

Characteristics of California farmworkers, 1992–1997. Source: National Agriculture Workers Survey.

By 1942, many farmworkers were drawn into the armed forces and industry, and growers fearing labor shortages persuaded the U.S. and Mexican governments to sign the first of what would become 22 years of *bracero* agreements that permitted Mexicans to enter the United States to work on farms. As rising incomes and population growth increased the demand for fruits and vegetables, and transportation improvements enabled California growers to produce commodities that could travel to the East Coast, the availability of *bracero* workers facilitated the expansion of agriculture. When the *bracero* program ended in 1964, many growers feared that lack of labor would force them to mechanize or stop growing labor-intensive crops.

UC received special funding to accelerate labor-saving research by redesigning plants and machines, as with processing tomatoes. Meanwhile, the absence of *bracero* workers enabled Cesar Chavez and the United Farm Workers (UFW) union to obtain a 40%, 1-year wage increases from some grape growers in 1966. There were predictions that the day of the unskilled farmworker was fast coming to a close, prompting the federal government to launch a series of programs that helped farmworkers, especially migrants and their children, to "escape" from farm work (Martin 1998).

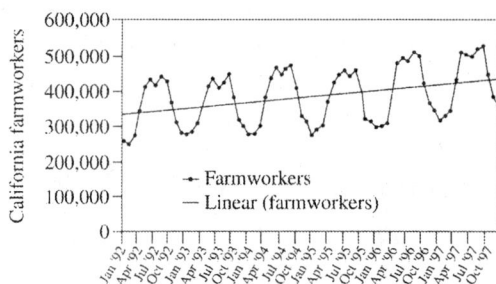

Farm Wages/Prices Index

Predictions of a mechanized agriculture proved premature. Americans increased their consumption of fruits and vegetables in the 1970s and 1980s, and Mexican workers continued to enter the state to do farm work legally and illegally. By the early 1980s, when the United States discussed imposing sanctions or fines on employers who knowingly hired unauthorized workers, farmers feared that immigration reforms would lead to labor shortages. The Immigration Reform and Control Act (IRCA) of 1986, which was intended to give agriculture a legal labor force and set in motion gradual wage increases, instead led to a new wave of authorized and unauthorized immigrant farmworkers (Martin et al. 1995).

IRCA created two legalization programs: a general program that granted legal status to 1.7 million illegal aliens who had resided continuously in the United States since Jan. 1, 1982, and the Special Agricultural Worker or SAW program, which granted legal status to 1.1 million illegal aliens who did at least 90 days of farm work in 1985–86; half of the SAWs legalized were in California. In addition, IRCA gave farmers two guest-worker programs under which they could obtain legal farmworkers if there were farm labor shortages.

According to the federal Commission on Agricultural Workers, appointed by the president and Congress to review the effects of immigration reforms on U.S. agriculture, the SAW program legalized about a million young Mexican men, equivalent to one-sixth of the adult men in rural Mexico in the mid-1980s (CAW 1992). The expectation was that these now legal immigrant farmworkers would continue to leave their families in Mexico, where the cost of living was lower, and commute seasonally between homes in Mexico and farm jobs in the United States. The fact that legal SAW farmworkers could take nonfarm jobs, it was thought, would force U.S. growers to increase wages and improve working conditions.

Both assumptions proved to be false. First, many of those legalized under the SAW program moved their families to the United States in the early 1990s. Second, farm wages and working conditions did not improve as expected because unauthorized workers continued to be readily available. Third, the farm labor market changed. As the percentage of unauthorized workers rose in the 1990s, the risks that an employer would be sanctioned for labor-law and immigration violations also increased. Farm labor contractors (FLCs) emerged as risk buffers between farmers and farmworkers. FLCs proved willing to assemble crews of workers and to assume the payroll and other risks associated with farm employment.

Farm labor contractors. Every year, hundreds of thousands of farmworkers are assembled into crews of 20 to 40 for jobs that typically last for a few weeks on a particular farm (Taylor and Thilmany 1993; Rural Migration News 1998). FLCs and foremen or crew bosses have been described as the glue that holds the farm labor market together, because they serve as intermediaries between workers and farmers.

In some cases, foremen or crew bosses are employed year-round by the farmer, and they recruit seasonal workers as needed. (In the scramble for workers, vans driven by *raiteros* asct as private transportation system in agricultural areas, ferrying workers between the farmworker sections of cities and fields.)

Each of the 1,300 FLCs registered in California in 1995 was required to obtain a license that costs $350 a year, post a $10,000 bond and pass a test on labor and pesticide laws (Rural Migration News 1999). Many FLCs are accused of taking advantage of vulnerable workers, levying unlawful charges for tools and rides to work, or not paying workers promised wages. Federal, state and local governments have erected an elaborate regulatory framework that attempts to encourage contractors and foremen to learn about and abide by labor and immigration laws, but there is considerable doubt about the efficacy of these laws. Between 1992 and 1995, a coordinated federal-state enforcement effort, the Targeted Industries Partnership Program (TIPP), found major violations committed by 90% of California FLCs inspected. A TIPP inspection of 23 FLCs with crews pruning vineyards in January and February of 1998 found that 52% of the FLCs were not paying their workers the minimum wage of $5.75 an hour (Rural Migration News 1999).

Unions. Unions have been active in California agriculture throughout the 20th Century, but most have proved to be short-lived. For example, the Industrial Workers of the World was active before World War II, the Cannery and Agricultural Workers Industrial Union was active in the early 1930s and the UFW has been active since the mid-1960s.

In 1975, California became the first major agricultural state to enact a farm-labor-relations law under which farmworkers could choose, under state oversight, whether they wanted to be represented by a union. If farmworkers voted for union representation in state-supervised elections, farm employers were legally obliged to bargain with the union the workers selected. The California Agricultural Labor Relations Board (ALRB) has supervised 1,600 elections on farms and certified 10 unions to represent farmworkers on about 800 farms since 1975 (Martin 1996). However, there were fewer than 300 union contracts in 1999, and about 200 of the contracts cover fewer than 10 workers each on the state's dairies.

The best-known farmworker union today is the UFW, which had a peak of 67,000 workers employed sometime during the year under 180 contracts in 1973. The UFW shrank to fewer than 10,000 members by the time Cesar Chavez died on April 23, 1993. Chavez was praised as the "Latino Martin Luther King," and was the 1994 recipient of a posthumous U.S. Medal of Freedom, the highest civilian award, presented by the president to honor those "who contribute significantly to the quality of American life."

Chavez's son-in-law, Arturo Rodriguez, became president of the UFW. In 1994, the UFW repeated its 1966 Delano-to-Sacramento march and announced that it would once again become active in the fields, organizing farmworkers, as it had done from the mid-1960s to the early 1980s. The UFW launched its campaign to organize strawberry workers in 1996, and targeted Coastal Berry, the largest strawberry grower in the United States. A competing union, the Coastal Berry Farmworker Committee, received 725 votes in a June 1999 election, versus 616 for the UFW.

It is not yet clear what impact the apparent Coastal Berry defeat will have on the resurgent UFW. Since 1994, the UFW has been certified as a bargaining representative for California farmworkers on 15 farms that involve a total of about 3,500 farmworkers. The UFW represents about 50% of cut-rose workers in the Central Valley and 70% of mushroom workers along the Central Coast.

In addition to the UFW, there has been a significant increase in the activities of self-help farmworker groups. As more migrants from southern Mexico and Guatemala arrive, there has been a proliferation of ethnic organizations, some of which have been recognized as unions by the ALRB. For example, the Mixtec and Zapotec Indians in California from the southern Mexican state of Oaxaca have formed "civic committees" in a number of California towns.

Guest workers. In the early 1980s, the percentage of unauthorized workers among California farmworkers was 20% to 25%, and farm wages and benefits were flat or declining. In the late 1990s, the percentage of unauthorized workers among California farmworkers was 40% to 50%, and farm wages and benefits flat or declining. Farmers in the early 1980s and the late 1990s feared a new round of immigration controls, and argued that before such controls could be implemented or improved, a new guest-worker program would be needed.

California farmers argue that they need an alternative to the 50-year-old H-2/H-2A program, which requires employers who want to have guest workers legally admitted to work for them to first receive a certification from the U.S. Department of Labor that U.S. workers are not available, and that the presence of the foreign workers will not adversely affect U.S. workers. Growers prefer a different admission procedure, called "attestation," under which the farmer attests or asserts that he tried and failed to find U.S. workers; this attestation serves as a permit to admit foreign workers. Enforcement would come after the workers arrived in the United States.

Unemployment data seem to belie the need for a new guest-worker program. Unemployment rates in the California cities in which many farmworkers live are very high, often 20% to 35%, even in September when farm employment peaks (fig. 3). With one in three workers unemployed even at the peak of the harvest season, and experience that "in the past, many temporary guest workers stayed permanently—and illegally—in this country," President Clinton on June 23,1995, issued a statement saying, "I oppose efforts in Congress to institute a new guest-worker or *bracero* program that seeks to bring thousands of foreign workers into the United States to provide temporary farm labor."

In July 1998, the U.S. Senate approved the Agricultural Job Opportunity Benefits and Security Act of 1998 (AgJOBS). AgJOBS would have substituted a registry run by the U.S. Employment Service for labor certification by DOL, and permitted farmers to obtain guest workers in an attestation-type procedure. Legally authorized farmworkers seeking farm jobs would have to register with local Employment Service offices. Growers would request workers from these registries and, if a farmer requested 100 workers and the register had only 50 available, the farmer would receive permission to have 50 guest workers admitted. Guest workers could stay up to 10 months in the United States, often shifting from one farm to another; if they did a certain number of days of farm work each year for 5 years—the suggestions are 90 to 150 days—they could earn an immigrant status under bills pending in Congress.

The current H-2A certification program is growing slowly. In 1997, DOL certified the need for 23,352 H-2A foreign farmworkers, up from 17,557 in 1996 and 12,173 in 1994. In 1997, 62% of the jobs certified were in Southeastern tobacco, another 18% were in Northeastern apples and 7% were in Western sheep herding, including California. Many of the H-2A sheepherders in California are from Peru, Mexico and China. Most are paid $700 to $750 a month and provided with a trailer and food. They usually receive 2 weeks paid vacation each year, and group health and worker's compensation insurance. Each shepherd is usually assigned about 800 sheep.

Future in Focus: No Major Changes Expected

One remarkable feature of the California farm labor market is how little change there has been in basic parameters over the past century—using bilingual middlemen to hire crews of seasonal workers, and worrying about whether enough workers will be available next year. A farmer from 1900 would be baffled by laser land-leveling, drip irrigation, vacuum cooling and the widespread use of computers, but would be very familiar with the use of bilingual contractors and crew bosses to assemble immigrant farmworkers to perform seasonal harvesting tasks.

The CAW final report (1992) called for an end to "agricultural exceptionalism," or special immigration and labor laws for agriculture; a renewed effort to reduce illegal immigration; and better enforcement of the labor laws that protect farmworkers. Six of the 11 CAW commissioners were from California. The commission surprised many observers by not recommending a new guest-worker program, instead calling for additional federal and state services for farmworkers, including more housing and services to assure equal opportunities for farmworker children.

How these workers and their children fare in their new communities will depend on government policy decisions, especially critical while the economy is strong.

Immigration policy is the wild card in shaping the future of the California farm labor market. If new entrants to the farm work force continue to be immigrants from abroad, then U.S. immigration policy will determine the number and characteristics of farm workers in the 21st century. Farmworker numbers and characteristics, in turn, will determine pressures for wage increases and benefit improvements. Immigration policy, a federal government decision, is the key variable affecting how immigrant workers and their children are likely to fare in California's rural and agricultural areas.

❏ References

California Assembly Committee on Agriculture. 1969. The California farm work force: A profile. Sacramento. April. 146 p.

[CAW] Commission on Agricultural Workers. 1992. Final Report. Washington, DC: Government Printing Office. 188 p.

Gabbard S., Mines R., Boccalandro B. 1994. Migrant farmworkers: Pursuing security in an unstable labor market. Washington, DC: US Department of Labor, ASP Research Report 5, May. 34 p.

Martin P. 1998. The endless debate: Immigration and US agriculture. In: Duignan P., Gann L. (eds.). *The Debate in the United States over Immigration.* Stanford, CA: Hoover Institution. p 79–101.

Martin P. 1996. *Promises to Keep: Collective Bargaining in California Agriculture.* Ames, IA: Iowa State University Press. 416 p.

Martin P., Huffman W., Emerson R., et al. 1995. Immigration reform and US agriculture. Berkeley, CA: Division of Agriculture and Natural Resources, Publication 3358. 580 p.

Mines R., Kearney M. 1982. The health of Tulare County farmworkers. Mimeo April. 46 p.

Rural Migration News. 1999. California: Enforcement, Workers Comp. 5(1), January. http://migration.ucdavis.edu.

Rural Migration News. 1998. Enforcement: Children, FLCs, MSPA. 4(3), July. http://migration.ucdavis.edu.

Taylor J., Thilmany D. 1993. Worker turnover, farm labor contractors and IRCA's impact on the California farm labor market. Am J Ag Econ 75(2):350–60.

[USDA] US Department of Agriculture. National Agricultural Statistics Service. 1975–1999. Quarterly. Farm Labor. http://usda.mannlib.cornell.edu/reports/nassr/other/pfl-bb.

[USDOL] US Department of Labor. Office of the Assistant Secretary for Policy. 1991–1998. National Agricultural Workers Survey (NAWS). Annual. www.dol.gov/dol/asp/public/programs/agworker/naws.htm.

BUILDING YOUR BUSINESS

Greg Thomas

Understanding how the local weather will create performance opportunities, combined with a cost-effective plan for acquiring customers, will steer you down the road to success.

Sadly, just going to the Affordable Comfort conference and learning the latest in building science, while fine-tuning one's technical skills, won't guarantee a successful building performance contracting business. In the 20 years that I have been working in this field, I have seen very technically talented building science believers lose a lot of money when they found out that consumers weren't knocking on their doors and their phones weren't ringing off the hook.

Yet there are building performance contractors who are running successful businesses. Just what are the successful business models for such contractors? Understanding the local market for comfort and energy improvements and knowing cost-effective ways to acquire customers are two essentials for creating a successful building performance business.

It's the Weather, Stupid

Success requires positioning yourself properly in your marketplace. To do that, you first have to understand how your local weather determines which building performance problems are most common in your area. Joe Lstiburek, principal investigator for Building Science Corporation, has identified different climate zones that affect building performance: Cold, Mixed, Hot-Humid, and Hot-Dry/Mixed-Dry. Each of these zones subject buildings to different combinations of temperature and humidity, and creates different kinds of performance problems.

A quick review of the opportunity list shows that HVAC problems tend to dominate in hot climates, and that envelope issues tend to dominate in cold climates. The more extreme the temperature or the humidity, the more severe the potential for performance problems. In truly extreme climates, many general contractors tend to know something about building science, because building failure is more problematic in these climates and the consequences can be so expensive. From Alaska to Saskatch-

---

ewan to Florida, liability driven by extreme weather has given building science businesses a head start.

Geography and the weather have also influenced construction styles, and therefore the opportunities to solve problems. In hot climates, attic ductwork has become the standard, creating its own set of problems caused by duct leakage. In the north, ductwork is usually in the basement. Hot-climate residential construction often doesn't even have basements. Sometimes really big problems get their start when an unsuspecting architect or developer imports a construction style, which may have been creating minor problems in one climate zone, into a more severe zone.

Some markets are less dependent on the weather. Multifamily markets tend to be driven by the cost of energy and increasingly by the liability associated with indoor air quality, such as second-hand smoke migrating through buildings. A market with a large number of FHA-financed homes may create the critical mass necessary for setting up a home energy rater and energy-efficient mortgage facilitator partnership. A rapidly growing market for new construction may drive home energy warrantees and Energy Star labeling. And health and safety considerations such as backdrafting are universal.

Who Ya Gonna Call?

After you have determined the best services to offer in your area, you have to figure out how to get potential customers to identify you as the solution to their problems, and how to do this in a way that is cost effective. Costs involved in acquiring customers include advertising and marketing expenses, the cost of time spent on the phone convincing customers to buy your services, the cost of calculating an estimate, the cost of writing up a proposal, and the cost of any other tasks you have to do to get a signed contract. If you get work from 50% of the potential customers you interact with, that work has to carry the cost of preparing proposals for the other 50% who chose not to work with you. In business terms, the cost of acquiring the customer has to be offset by the gross margin on each job you sell. The smaller the job, the more jobs you need to sell, and the more you need to have a very low cost means of acquiring customers.

The high cost of educating customers so that they will spend the extra money on the contractor with the diagnostic skills has been one of the biggest barriers to the growth of this industry. But this cost is unavoidable, no matter what type of building performance contractor you are or plan to be (see "Fitting into the Marketplace"). Building performance contractors can't compete on price against contractors who don't buy diagnostic equipment or attend training events.

The cost of acquiring customers is a particular problem for energy raters. If a rater must sell each customer, one by one, on the importance of getting a rating, the cost of acquiring customers is too high relative to the low gross margin on each job. To survive, most successful energy raters employ one of three strategies aimed at getting someone else to bear the costs of acquiring customers: They set up large referral networks with banks, they maintain quality assurance relationships with many builders, or they get work from energy-efficient mortgage facilitators who have themselves created the necessary referral networks.

Instead of relying solely on diagnostics to bring in revenues, many building performance specialists combine contracting with diagnostics. This strategy increases the gross margin per job and reduces the need to acquire so many customers. Combining contracting with diagnostics can also be a strong selling point for consumers. Consumers would prefer to find a one-stop shop, because it saves them time and allows them to establish trust with one contractor instead of several. If you provide only the diagnostic half of the solution and leave customers to their own devices for the other half, will they see your service as useful?

Cost Effectively Getting Customers

Comfort Diagnostics in Little Rock, Arkansas, started out just doing diagnostics on buildings, but quickly found out that customers like to pay for work, not for opinions. In a move to combine doing inspections with contracting, they started offering insulation services. When they realized that they were subcontracting out too much HVAC work, they acquired an HVAC contractors license. This enabled them to do large jobs with larger margins. As a differentiated high-quality HVAC contractor, they also had an identity that consumers could recognize and easily locate in the yellow page listings.

Comfort Diagnostics' next challenge was to acquire more customers. Blessed with glib tongues, they turned to radio, and invested heavily in an hour-long talk show that focused on using building science to solve comfort problems and reduce high utility bills. Now the talk show, combined with the resulting word-of-mouth referrals, brings in the bulk of their business. The investment in radio was not cheap; it required a long-term commitment of time and money. But it has paid off in an ongoing stream of customers. Radio is an excellent educational medium, and what Comfort Diagnostics is doing is educating customers through a weekly repetition of practical building science advice and testimonials from customers. Thanks to their rapid adaptability, Comfort Diagnostics has the fastest growing building performance business that I know about.

Fred Lugano of Lake Construction is another successful building performance contractor who combines diagnostics with roofing and remodeling. Located in Charlotte, Vermont, Fred realized that rotting roofs and ice dams were a big potential market for his services. But instead of hanging out a shingle as a building performance contractor, Fred invested in yellow page ads in the "Roofing Contractor" section.

These ads focused on Lake Construction's ability to solve a variety of roofing-related problems. As Fred puts it, "Sell the benefit, not the service."

Another means of acquiring customers is through contractor-to-contractor referral systems. In Toronto, when a roofer sees a moisture related problem, he or she calls in a building performance specialist and includes the building performance work as a subcontract in the roofing proposal. To provide customer service benefits, the utilities in that area are now sponsoring the development of a trade association to help support this referral process. There are many of these referral networks, generally on a smaller scale, in New York and elsewhere. IAQ referrals are also gaining in numbers. Home performance contractors are linking with healthy indoor air groups and are getting a steady stream of referrals.

Still another profitable type of partnership is a linkup between an HVAC contractor and an insulation contractor. Ed Bishop of Advanced Heating Concepts and Pete Collisson of Affordable Energy, who met at a utility training for whole-house contractors, and who operate outside of Albany, New York, teamed up to share leads and subcontracts. Ed does the HVAC design and installation, while Pete does the diagnostics and the envelope work. The combination lets them offer the performance and cost benefits of a whole house approach. When one sees a job that needs the other's skills, he gets a quote for a subcontract.

Business Development Resources

Once you have a handle on the motivating factors in your local marketplace, and a sense of how you might acquire customers, examine your skills and seek out training in the areas where you are weakest. To learn more about running an independent contracting business, consider enrolling in either of the National Association of Remodelers (NARI) or the National Association of Home Builders (NAHB) Remodelers Council certification programs that cover mostly business issues. For less time-consuming tips on improving an existing operation, I have found the Small *Business Advancement National Center Newsletter* to be a helpful resource. To receive this newsletter, send an e-mail to webmaster@www.sbaer.uca. edu with the word "add" in the subject line.

The small-business side of being a HVAC contractor has been explored and well documented by the Air Conditioning Contractors of America (ACCA). They have many resources supporting the development of a professional, quality-driven HVAC business. In addition to business resources, ACCA publishes a number of technical manuals for residential and commercial HVAC design. For example, ACCA's *Manual J* is the reference for residential system sizing, and *Manual D* is the reference for residential duct design. I find the current version of *Manual D* useful when talking to contractors because it is an HVAC industry source that references duct leakage studies and blower door testing. All of these organizations have excellent Web sites.

Another excellent reference for local training opportunities is *Home Energy's* training guide ("Training for Tomorrow: Are Your Contractors Certifiable?" *HE*, Jan/Feb '99, p. 29). This guide is also available on the Web. *(Home Energy's* Web site is generally a great place to get information, and I often tell potential customers to visit it.)* I also use the *Home Performance Contractor* brochure and the *No Regrets Remodeling* book, both published by *Home Energy,* as third-party educational materials for consumers.

Sales: Open with Comfort and Close with Efficiency

Whether you choose the company route or the independent-consultant route, you must be comfortable with the sales process. The sales process is about building trust. Use your building science knowledge to initiate the process, and use third-party sources of information—such as Energy Star program materials, the *Home Performance Contractor* brochure, or reprints from *Home Energy*—to support your assertions and consolidate your relationship with your customer.

Despite all the public funding spent to promote energy efficiency, most homeowners still equate saving energy with living in a cold, dark house. Rather than trying to push energy savings "predictions" on an unwilling customer, try emphasizing the non-energy benefits of building science. Customers have seen predictions come and go, and it will be very hard to make them believe in yours. But by documenting the symptoms of building failure and providing clear explanations and third party support for your explanations, it is very possible to convert a customer into a believer in building science and performance testing. Then you hit them with the close: "And it saves energy and pays for itself!"

Shopping for Solutions

I have used this strategy numerous times to convince potential customers to sign a contract with me. In one such case, a condominium association that had experienced $80,000 in ice damage over a five year period was desperately looking for a solution. They had hired architects and engineers. They had put in added insulation and attic venting. Nothing had worked. Finally, their insurance was canceled because of the recurring ice damage. Through a utility referral process, they contacted a couple of local building performance contractors that I team up with from time to time.

Working with one of the contractors, I went into the building and took pictures of all the usual suspects: attic duct leaks, drop ceilings, pocket doors, lighting fixtures, plumbing chases, and so on. Also, we found that the gas vent for the downstairs apartments was actually run through the boxed-in cold-air return of the upstairs apartments. We went back and worked up our estimate. There was a time when I would

Fitting into the Market Place

Contractors can be divided into three groups based on how they use building science and performance testing in their business. These categories make it easier to target building performance marketing and training programs to reach specific kinds of contractors.

The Building Performance Contractor

The highly differentiated, problem solver contractor is the one customers yearn for when they have a nasty building performance problem that nobody else can solve, or when they really want to ensure a high-performance building. Unfortunately, customers have to seek high and low to find such a contractor, and in many areas no such contractor exists. Start the search by contacting organizations that fund contractor trainings and related programs, such as utility companies, weatherization agencies, energy rating organizations, and state energy offices. This type of contractor lives by his or her reputation and depends on a strong referral network.

The Top-Quality Trade Contractor

The second type of contractor integrates building performance into his or her existing trade, be it HVAC, insulation, roofing, remodeling, or new construction. This contractor uses building science and performance testing to improve the quality and scope of the work that he or she already does. The real distinction between this type of contractor and the building performance contractor is in how they get their customers. Customers can find the second type of contractor using listings in the yellow pages for the already-established trades. Once the customer has made contact, these contractors use building performance techniques to differentiate themselves from other contractors. Because they offer a recognized service, they can transform customers who call them for roofing, for example, into customers for building performance services.

Top-Quality Builders and Remodelers

Builders and remodelers typically do not want to train their own crews to undertake specialized tasks, especially ones that require a high level of quality control. It is easier to hire a subcontractor. For builders and remodelers who are educated on the benefits of building science and performance testing, a local subcontractor with building performance skills is worth his or her weight in gold (OK, maybe dense-pack cellulose). As Energy Star and green building programs get more public recognition, these builders will require more and more building performance subcontractors.

have written a long report that included a sophisticated savings estimate, but I stopped doing that when I discovered that people want to pay for the work, not for a long report. The summary report I gave to the condominium association was mostly a bulleted list of the problem areas and an attachment of photos mounted on card stock with nice captions.

We made our presentation, and passed around the photos. We answered a few questions about the ice on the buildings, and then they asked, "Will we save money?" I said, "Yes," and the job was ours. No time spent on savings calculations. (We did look at the billing data to score the energy intensity of the buildings in Btu/ft^2 per heating degree day.) Still, they were not ready to buy until there was an economic rationalization, however weak, for doing so. They didn't call to save energy dollars, but saving energy dollars made the investment smart and politically safe. The same logic applies to people making decisions at home. We make them feel smart when we combine the motivating creature comforts of satisfactory indoor temperatures, good indoor air quality, and durable buildings with the economic rationalization of energy savings.

Untapped Market Exists

I firmly believe that there will eventually be a long term consumer driven market for applied building science and performance testing. Studies by Consumers Gas in Toronto and by *Contracting Business* magazine in the early '90s found that 10% or so of homes have building performance-related problems that their owners want to solve now. There should be plenty of business in that 10% of homes to keep many home performance contractors very busy.

Furthermore, as Energy Star and green building programs get more public recognition, more and more builders will require the services of building performance subcontractors. I witnessed this phenomenon firsthand at the first National Green Building Conference put on by the NAHB Research Center. There I heard builder after builder get up and say that he or she had gone green and was pleased to be both doing the right thing and making more money by selling more houses more quickly. These

---

Training and Certification

In 1979, I put my engineering background to work helping a friend who was starting a business installing solar hot water heaters and wood boilers. While researching how to save energy, I came across information on blower doors and started reading the ASHRAE transactions on measuring air leakage in buildings. I built a blower door and began offering "Whole House Audits" as a way to differentiate our heating company from other contractors. We created a relationship with an insulation contractor to offer a complete HVAC and air sealing package.

After a few years of selling and installing furnaces and doing air sealing, I ran into Larry Kinney, of Synertech Systems, who was building a blower door and doing DOE low income weatherization training and technical assistance work. I ended up working for Larry and was the lead trainer for introducing blower doors into the Pennsylvania and New York weatherization programs.

In 1986, I went off to restart a dormant not-for-profit energy service company, Syracuse Energy or SyrESCO. We did a lot of work with not-for-profits and multifamily buildings. During this time, I also got my heating contractors license. In 1993 when I left, there were over 20 employees. Since leaving the not-for-profit, I have completed a myriad of energy efficiency consulting projects, including program, software, curriculum, and association development.

builders' big problem now was finding qualified insulation, HVAC, and blower door testing subcontractors who could do the building performance work required to make their green buildings function well. Some of these builders had been driven to the extreme of capitalizing and training new subcontractors because their existing subs were unwilling to make the switch. Now that's dedication, and it offers a real opportunity for subcontractors with building performance skills.

There is just too much value in the services we offer for the market to ignore them much longer. The ability to take control of the movement of air, the transport of moisture, and the transfer of heat in buildings allows us to provide high-quality, healthy living environments—and we save energy for our clients. With an ever-increasing concern for health and comfort, how can we lose? Those people who combine building performance expertise with good business skills will have a head start in mining or creating that market in their own backyard. Good luck!

THE GOOD DADDY

Suzanne Braun Levine

Today's employers call themselves family-friendly, and for women, that's increasingly true. But men are still expected to make work their top priority. Here, the stories of a few courageous fathers who are risking their careers to stand up for their families—and overturn society's double standard

Matt, a thirty-eight-year-old father of twins and an attorney at a small law firm, has become a worry to his boss, Jeff. Although Matt works hard, "he just takes off in the afternoons, like to go to his three-year-olds' birthday party," says Jeff, a senior partner who prides himself on having created a family-friendly environment, where no one works weekends or stays late. "Why can't he and his wife schedule these things for five o'clock instead of three?" When asked whether he would resent the same behavior on the part of a female employee, Jeff acknowledges that he has different expectations for a working father than for a working mother.

He's not alone. Employers say they are becoming more family-friendly; in fact, according to a recent report by the Families and Work Institute, a New York City–based organization that studies workplace issues, some 56 percent of 1,057 companies say that they offer "employee assistance programs to help workers deal with problems that may affect their personal lives." But men are less likely than women to avail themselves of these opportunities. Those in the know pass the word that it's better to call in sick than to take a "family day." Or, when a child is born, it's smarter to take vacation days than paternity leave.

Surely employed fathers are entitled to the same options as working mothers. The few men and countless women who take advantage of flextime, job sharing and other benefits know that their decision may slow their careers. But women can count on respect, if not support, for their choices; men cannot.

For a man, it's still professionally risky to behave in any way that suggests that work is not the top priority. This puts the working father in a dark and lonely place that's difficult to escape from because it's solitary confinement. There aren't many accounts of life behind the wall of silence, where men keep their real priorities, because anyone who wants to be a father "on company time" quickly learns that he can do so only by flying below the radar. The good news is that there are more and more dedicated dads working and living in a wide range of circumstances.

James Levine, founder and director of the Fatherhood Project, an organization in New York City that tracks the experiences of fathers at work, has become an expert in luring them out of the woodwork. Increasingly, he is being asked to advise employers who want to convince mistrustful employees that they're committed to family concerns. The first thing Levine does at a company is to announce a meeting of fathers to discuss work/family conflicts. The response is enlightening. Many of Levine's clients have little sense of how the fathers in their organizations are struggling until they learn that the meeting is oversubscribed within hours. And the fathers are just as surprised by the turnout; each thought he was alone in his dilemma. Not only is he not alone, he is part of a revolution—the same revolution that brought so many women into the workplace. The problems today's fathers confront are, in fact, part of a larger political, social and economic landscape that is limited by outdated expectations for men and women and tilted against the needs and realities of family life.

All Work, All the Time

Giant strides in telecommunications, once touted as creating more time for family, have instead made the "24-7" (twenty-four hours, seven days) workweek possible. I recently heard a young, energetic investment analyst rave about his electronic workplace: What appeals to him most is that, since the world economy is always open for business, he can stay plugged in every day—even from home. "With all the new technology," he announced, "I can now spend more of my weekends with my family!" His family may have a different opinion, though, as he plunks away on his laptop and juggles overseas conference calls on weekends.

Another source of family disruption is business travel, a requirement of a growing number of jobs. Ralph works for a major computer publishing company, where dress is casual and the policy is family-friendly. Nonetheless, the business is international and the travel unrelenting. Last year he was in Sri Lanka on his daughter Tillie's fourth birthday; she was so disappointed that he promised to be at her fifth birthday party, no matter what. Ralph alerted his boss to that commitment months in advance. Then something came up—an important management conference. Ralph reminded his boss that he couldn't go. His boss was sympathetic (he has young kids, too) but insistent. "Why don't we send your family?" he offered. "Wouldn't that be a great birthday present?" Well, not really. Not to a five-year-old who wanted a real birthday party, with favors and friends and cake *and* Daddy.

Over the next few weeks, tension between Ralph and his boss rose as they tried to resolve the impasse. The story has a happy ending: A subordinate went to the conference, and Ralph attended the birthday party. But the amount of energy and conflict that went into settling the question of a single birthday party testifies to the intransigence of business travel. Not surprisingly, all the women at Ralph's company had opted out of his travel-intensive career track.

Taking a Stand

Ralph was lucky. Typically, when good intentions conflict with business considerations, work wins out. That's what Conrad, a human-rights activist, discovered when he took a job with the understanding that "I will work forty-five hours a week, and you will get superlative effort and ability in those hours." Four years later, he's concluded that "the deal was bull."

As he sees it, "They wanted to hire me, so they said, 'Sure, we're a family-friendly workplace. Nobody works crazy hours here.'" But as the organization grew and prospered, more traditional measures of productivity clicked in. "If somebody is prepared to give you fourteen hours a day and someone else will give you nine, you're going to take the fourteen."

Conrad also noticed another kind of bull—the fourteen-hour day disguised as nine: "My boss has two kids about the same age as mine. He doesn't stay late. But at eight-thirty the next morning, he's got a stack of memos and letters to prove that he works till midnight every night."

Conrad tries to "do unto others" when hiring subordinates. "The first thing I say is, 'Don't look at your watch at five thirty-one to see if I'm noticing that you're still here. If you need to go home, go home.' But in terms of stamping that philosophy on the institution, forget it."

If Conrad, the human-rights activist, isn't ready to take on the system, who is?

Carl is an intriguing example of someone who knows how to play the duck-and-feint game. As a human-resources manager, he sees family pressures both professionally and personally. When his son was born, Carl's priorities shifted. Work and money took a backseat to being home with his son. "If that means I'm not going to get as high up the chain as I would have otherwise, so be it," he says.

Does he take advantage of the family-friendly programs that he offers other employees? Not really. Nor do the men he counsels. "I've yet to hear a man say 'I want to work from home to care for the kids,'" he says.

Carl has been able to tailor his schedule to family obligations without appearing to. Because he travels to his organization's outlying branches, he has some flexibility—and some invisibility. "On Friday, I'm at a site that's five minutes from my house. I have told my assistant to avoid scheduling meetings after three o'clock, so there's less chance that I'll be late for what I call my second job, my six o'clock commitment to my son."

Carl had to attain seniority to be able to design such a schedule and even so, he admits, "I definitely look over my shoulder as I'm walking out the door at five or five-thirty, because everyone else is still working."

Well, not *everyone*. It only looks that way to the guilty few. "There are some people in similar situations," Carl concedes. "We're like Alcoholics Anonymous; you never identify who you are, but everyone knows you."

If Carl, the human-resources manager, won't take advantage of family-friendly policies, who will?

Stepping Off the Fast Track

Nick has created his own family-friendly work environment without benefit of company policy or subterfuge. His blend of family and work commitments is typical of a working mother's. "You just have to be ultra-efficient," he explains. "I never stop working throughout the day. I eat lunch at my desk and answer my E-mail." And when Nick has squeezed the maximum work into the given time frame, he leaves. Nick empathizes with women on the so-called Mommy Track. "I understand that you can't be at the top—at least for a time."

Some men, unable to incorporate family time into their workdays, are simply forced to change jobs or careers. By the time Brian, thirty-six, graduated from college, he was married and a father. Two year's later, he set out on a fast-paced career in finance. When his son and daughter were little, he worked long days and weekends. That went on for ten years—until he got a call from the guidance counselor at his son's school. "He said, 'I'm really fond of your son, but he's starting to hang out with kids who are having behavioral problems.'"

"I was kind of shocked," Brian admits. "I said, 'What can I do?' And he gave me the best piece of fatherly advice I've ever gotten: 'Spend time with your son.'" To do that, Brian knew he had to change his work life. It wasn't easy to find a more forgiving job, but he finally ended up at a very low-key financial operation run by an older and understanding friend.

Brian's relationship with his son didn't improve overnight. "Billy was resentful and angry that I had ignored him," he says. Brian's family had to make financial sacrifices, such as staying in a small apartment while their friends were buying houses. But Brian has come to see his dream of having it all in sequential terms. "I was young enough to say, I'll spend five years with the kids and then I can pursue my next career."

Changing the Rules

Slowly, employers are being forced to recognize the needs of the working father. According to Ellen Galinsky, president of the Families and Work Institute, "When we first started doing focus groups, the men and women sounded very different. If the men complained at all about long hours, they complained about their wives' complaints. Now, if the timbre of the voices were disguised, I wouldn't be able to tell which is which. The men are saying: 'I don't want to live this way. I want to be with my kids.' I think the corporate culture will have to begin to respond to that." But first, the voices in Galinsky's focus groups need to go public.

Where are the men who are willing to stand up for our families the way their forefathers stood up for our country? They're out there, and they will find each other—if not in protest marches, then in the quick-getaway corners of company parking lots. As they stand up, they will be counted. And leaders will emerge—if not because they have the courage to speak out, then because they have the power to change the rules.

AFFORDABLE RURAL TRANSPORT

Priyanthi Fernando

Development workers often fail to recognize the importance of transport in rural development. Rural development organizations rarely see themselves as dealing with rural transport issues, and the transport sector has conventionally focused on large infrastructure projects and is only just beginning even to consider the transport needs of rural women and men.

The consequences of this neglect are considerable. Robert Chambers has identified isolation as one of the key elements of poverty.[1] Isolated communities have little or no access to goods and services, and few opportunities to move outside of their immediate physical environment. This has been shown to restrict their agricultural productivity, reduce their health and educational status, and limit opportunities for employment and for political participation.

The British Department for International Development (DFID) is encouraging the use of a "sustainable livelihoods" approach as a coherent and consistent strategy for eliminating poverty in the poorer countries of the world.[2] This approach provides a framework for analyzing individual, household, and community needs as a basis for targeting poor and vulnerable people. The approach centres around peoples' livelihood assets (human, natural, physical, social, and financial) the structures and mechanisms that constrain their access to these assets, and the livelihood strategies that can overcome these constraints and enhance their livelihood opportunities. Transport provision plays many roles in this complex arrangement. On the one hand, transport contributes to the level of physical assets (infrastructure and means of transport) available to a community, household, or individual. On the other the lack of it undermines the capacity of communities, households, or individuals to access, or draw on, their other assets. Restricted access, for instance to education and health services, can impair peoples' ability to make the best use of their human assets.

The burden of transport activities puts severe pressure on the available time of poor individuals and households. Basic household tasks, such as collecting water and firewood and travelling to and from the fields, are both time- and energy-consuming, particularly for women. Poor women, men, and children spend so much time meeting subsistence needs that they lose out on opportunities to earn higher incomes and to develop their financial resources. Then, because they have limited financial resources, they are unable to invest in means of transport or in transport infrastructure.

The type of analysis that the sustainable livelihoods approach brings to development helps us to realize the central role that the lack of appropriate transport plays in maintaining the cycle of poverty and isolation and to pinpoint the need to develop alternative, more affordable transport systems for the poor. Developing such systems requires consideration of four key elements. They are:

❑ the improvement of village-level infrastructure such as paths, tracks, and footbridges;
❑ the promotion of the use of intermediate means of transport (i.e. means of transport that fall between walking and high-cost motorized vehicles such as cars and trucks);
❑ the provision of adequate and affordable rural transport services; and
❑ the siting of services closer to the communities, thereby removing or reducing the need for lengthy travel. (This component is sometimes called a "non-transport" component because it does not involve interventions that relate either to the means of transport or to infrastructure.)

Members of the International Forum for Rural Transport and Development (IFRTD) and organizations such as Intermediate Technology, the International Labour Organization (ILO), parts of the World Bank, DFID, other bilateral agencies such as NORAD, SIDA, and DANIDA, and others, all advocate this broad approach to meeting the access and mobility needs of rural women, men, and children. The ILO has also developed an alternative, participatory planning methodology that helps planners to identify household accessibility needs and the options for meeting them at a decentralized level. Called "integrated rural accessibility planning" (IRAP), the approach has been piloted in Malawi, Tanzania, the Philippines, and Laos and has led, among other things, to a greater consciousness of the need to measure the value of time, particularly time spent on domestic tasks with no immediate economic value.

In this issue, we look at a range of interventions that have improved the affordability of transport for the rural poor. Affordability means a number of things. In the first instance it means the availability of appropriate low-cost interventions, Most commonly this refers to a range of low-cost, intermediate means of transport (IMTs) that can be made are available to poor people in rural areas in Asia, Africa, and Latin America. Almost always this range includes animals and animal-drawn means of transport. Donkeys in particular are owned and used by poorer households, usually small farmer households, pastoralists, and micro-entrepreneurs in the transport sector in developing countries. They are no longer prevalent among large-scale commercial farmers or in commercial operations such as mining. In most countries donkeys are cheaper than other work animals because they are not perceived as multi-use animals (cattle, buffaloes and camels are usually kept for their milk and their meat as well as for work), nor sold for their meat, nor considered in the payment of bride price. This makes them more affordable to small farmers. Other low-cost means of transport include different kinds of pack animals, sledges, animal carts, bicycles, and other cycle-based transport modes, and (at the upper end of the range) cheaper motorized means such as mopeds and motorcycles.

Alternative Means

The use of these IMTs can lower transport costs enormously and they have the potential to increase the efficiency of on-farm agricultural transport. They can allow faster travelling and the transport of larger loads. If used for domestic tasks such as firewood and water collection, they can save a great deal of time, particularly for women. They can also improve access to public services. Few of the poor in rural areas can afford any of these means of transport, however. Studies in Africa have shown that bicycle costs can range from 20 to over 100 per cent of a rural household's annual income.[3] A study in Zimbabwe has shown that only wealthier families own IMTs such as scotch carts.[4] Women, who in most societies have less access to disposable assets, are also less likely to be able to afford intermediate means of transport. Fewer women than men ride bicycles, use ox carts or wheelbarrows, or hire transport services. Women's inability to afford means of transport is often exacerbated by the social and cultural factors that define gender power relations, and determine what women can and cannot do in different societies. Researchers from Bangladesh have argued that mobility ought to be considered a human right, since restrictions on mobility prevent women from accessing existing opportunities for education, employment, and political participation.[5]

Making intermediate means of transport more "affordable" to poor rural women and men may require looking beyond the basic costs of the technology to providing credit (or subsidies) for their purchase—which can stimulate the adoption of IMTs. In the Tanga Animal Draft Power Project in northeastern Tanzania in 1981, lack of farm power and transport were identified as crucial constraints for smallholder farmers. The farmers wanted the donkey carts that were designed and tested by the project, but could not really afford them. This was partly solved by a labour-intensive rural road maintenance programme that contracted farmers to bring gravel to resurface the roads. The income from this programme made it possible for farmers to repay the loans, and the carts were used for many purposes other than gravel haulage. There are also examples where credit did not provide a sufficient incentive to lead to wider adoption. In Ghana, the Northern State Feeder Roads Program sponsored by the World Bank set up a similar saving scheme for the purchase of cycle trailers with deductions at source for those working on the labour-based road works. But the trailers were not popular. They were expensive relative to local incomes, and many of the women who worked on the programme did not own bicycles. The trailers were also intended for village-to-market transport on the new roads, but market trucks were now available and widely used. Most importantly, the ordinary bicycle was considered a very flexible transportation device capable of carrying significant loads and which was, at half the price of the bicycle/trailer combination, much more affordable![6]

Credit programmes usually require borrowers to have a source of income that will ensure repayment of the loans. Women (especially African women) spend more time on transport activities than men, work longer hours than men, and in most developing countries carry an average of 53 per cent of the total burden of work.[7] Much of women's transport burden is, however, spent on what gender analysts would call "reproductive" work, or domestic chores such as water and firewood collection and food (not cash crop) production. It is obvious that the use of intermediate means of transport could relieve this burden on women. What is less obvious is how women with little or no cash income could afford them and how credit schemes could be developed to address their need (particularly since in many instances it is the time taken for transport that restricts their engagement in additional income-generating activities).

Low-cost intermediate means of transport need to be supported by an affordable system of manufacture, supply, and repair. Several projects have attempted to boost local economies by encouraging artisanal production of IMTs at a village level. The Department of Engineering's Development Technology Unit (DTU) at the University of' Warwick, UK, has been developing low-cost animal cart technology in Nigeria, Kenya, and Uganda. Working with artisans in their own workshops with existing tooling has lowered costs by 50 per cent in some localities,[8] but artisanal production may not always be the most effective way of providing affordable transport technologies on a large scale. Small workshops often cannot effect the economies of scale that result in low-cost products. The success of SISMAR carts in West Africa is probably a result of their centralized production in Senegal. IT Sri Lanka's cycle trailer project could have produced cycle trailers at a lower cost and higher quality and thereby benefited a greater number of users, had they been produced commercially, but the project had made a conscious decision to go with artisanal production.[9] In the five years since the project began, it has disseminated only about 500 to 600 trailers in a country with an estimated two million bicycles.

Spares and Repairs

Affordability is also related to the availability of affordable spare parts and repair services. Usually for such support to be available there needs to be a critical mass of the means of transport in the area.[6] Starkey describes the situation in the village of Anjanadoria, 70km from Antananarivo in Madagascar, where most families own an ox cart, there are about 800 carts in use, and there are two carpenters who make and repair ox carts in the village. But there is no-one in the village who repairs bicycles. Bicycles are taken by ox cart 15km to the local market town to be repaired, and the absence of local repair facilities are one reason why so few people own bicycles. At the same time no one had started cycle repair services in the village because there are so few bicycles and little demand.[6]

A serious, yet less well-recognized issue, is that of the character and type of engineering used in developing countries and who benefits from this. Engineering detail design is highly political, and increasingly manufacturers are using "repair proof" designs to control maintenance (and thus product life) and to acquire market share of the repair and spares market (from which they derive most of their profit). Designing affordable transport technologies for the poor in developing countries requires an alternative approach. Cost-effective transport for those with minimum resources depends on the technologies being available in a usable condition. This implies that the technologies need to be designed for repair in the user's environment from the outset, and that the mean time to repair ought to replace mean time to failure as a measure of engineering quality.[8]

The mobility and access needs of rural women, men, and children can also be met through both sharing and hiring arrangements for the modes of transport available and the provision of affordable transport services. Transport services are of particular importance to people who do not own their own modes of transport and who would otherwise have to travel by foot. Even though much rural transport takes place within and around the villages, rural people need to travel outside their immediate environment to school and health clinics (where these are not available in the villages), to carry goods to market, to visit friends and relatives, and to engage in social and religious activities. In most developing countries, the poor condition of rural roads, the high cost of vehicles, restrictions on their supply, the non-availability of spare parts, inadequate maintenance capacity and capability, and the loss-making nature of public transport services constrains the provision of affordable rural transport services to these people. Most commercially operated services concentrate on longer distance main routes that have the best operating conditions and the highest transport demand. Local systems are unreliable, unsafe, and unusable with consequences for the social development of women and men living in those areas. In Pakistan, where female literacy is only 23 per cent, women surveyed in the Multan, north Sindh, and Islamabad areas said that their families would not allow them to attend school because they had to use overcrowded means of public transport.[10] In this issue we describe two attempts to provide affordable transport services to people in the rural areas of Eastern Africa (Uganda and Kenya) and Sri Lanka.

The Community Bus Service initiated by the Lanka Forum for Rural Transport Development and villagers in the south-eastern district of Ratnapura is an attempt to fill the gap in transport provision created by the privatization of bus services in Sri Lanka. In most parts of the world, private companies have taken over the provision of transport services previously provided as a public service by the state. In several developed countries, however, there are examples of governments using a needs-based approach in the transport sector for special groups, especially those experiencing greater deprivation. The UK government has subsidized the capital and operating costs of shipping and air services to the more remote islands of Scotland, and countries such as Norway, Switzerland, and the USA have special programmes of assistance for low-density regions such as farmers in the mountainous regions. But, as John Howe has pointed out, many aid agencies and financial institutions seem reluctant to treat transport provision as other than a hard economic good, contradicting the behaviour in their own countries.[11] In Sri Lanka, privatization has been implemented in the interests of efficiency and is consistent with the economic structural adjustment programmes advocated by donors. But the Government of Sri Lanka, mainly for political reasons, controls the price of bus fares. Low fares combined with the poor state of infrastructure has severely lowered the level of transport services to those people in areas that are deemed "unprofitable" by private bus operators. The state provides a small subsidy to those companies able (and willing) to run on these routes, but because the subsidy fails to cover the costs, the service is irregular. The community bus service aimed to overcome these problems by providing the villagers with a transport service which both meets their needs and over which they have control. From the point of view of the policymakers in the Lanka Forum for Rural Transport Development, the service provides an "experiment" of an alternative way of providing transport services to rural areas. Such experiments have taken place in other countries, but they have been short-lived and most transport professionals are skeptical about the sustainability of the service. Whatever the outcome, the "experiment" should generate some interesting lessons.

Entrepreneurs

In East Africa, the development of *boda boda* bicycle and motorcycle taxis are a spontaneous, entrepreneurial response to freer market conditions and the availability, as a result, of greater numbers of bicycles and motorcycles. *Boda boda* operators are mostly young men who provide a taxi service between rural areas and rural towns in many districts in Uganda and (more recently) in Kenya. Users of these services include people who work outside their villages, the business community, students, patients going for treatment, and—not quite the target audience imagined—thieves in the night.[12] *Boda boda* operators are usually quite well organized, and the service, while providing an affordable transport service to the users, is also an income-generation activity for the men who ride the bicycles/motorcycles. The benefit from *boda boda* transport for women is, however, limited, since fewer women than men work outside their villages or engage in transport activities that take them to the rural towns. Most of a woman's transport burden results from travel within the village or nearby—a burden that is not alleviated by the provision of affordable transport services.

The Right Technology

The idea of affordability has in its most obvious form been applied to the capacity of individuals to own means of transport or to use transport services. But often communities (or even nations) cannot afford the needed stock of rural transport infrastructure. As with vehicles, this is often a result of the cost of the technologies used. There has been a tendency in the past to over-design roads and to use very capital-intensive construction methods. Labour-based techniques have been proved to be more cost effective and to add value to local economies by creating employment and jobs. Inherent in the low-cost, affordable road which serves most communities in developing countries is the crucial need for maintenance.[13] But maintenance requires resources and planning. Financing needs can be reduced by lowering technical standards, choosing more appropriate technologies, and changing the mode of procurement, or by considering different institutional options such as cost sharing with local road cooperatives.[14] Funds for maintenance come from local government taxes or from central government transfers. Unfortunately, such monies are rarely adequate. Several countries have some success in making road users pay directly for the use of the roads. This has been implemented in the form of a commercially managed road fund at the national level. Road fund revenues come primarily from vehicle license fees and a levy added to the price of fuel. The most common method of allocating resources to rural road maintenance has been on a percentage basis, where 25 to 35 per cent of the road fund is allocated to the rural road network.[14]

There is an increasing interest in the transport sector about community participation in road maintenance activities as a way of ensuring the sustainability of community infrastructure. The assumption is that if communities feel a sense of "ownership" of the road they are likely to invest time, energy, and material resources to maintain it. A more cynical approach to community participation argues that it is merely a way of extracting resources from the resource poor in the context of decreasing public spending by the state. Ideally, community participation should be about empowerment, but communities usually participate in projects initiated by development agencies and not the other way round. Some advocates of community participation make the assumption that rural women and men are completely disempowered and ignore the spontaneous efforts of communities to initiate their own projects. But there are examples of communities in developing countries who have voluntarily banded together to construct roads and bridges. The Sri Lankan villagers building the road to run their community bus were able to manipulate a range of local government institutions to divert resources in their direction. Another group were able to overcome many, seemingly insurmountable, technical problems to build a road leading to their village.

Transport professionals are beginning to accept the significance of rural transport in the livelihood strategies of poor women and men in the rural areas of developing countries. As their focus shifts from large-scale infrastructure projects to affordable interventions for the poor, it is important that they recognize the capacity of communities and community organizations working at village level. At the same time it is important for people and organizations working with communities in rural development to address the transport needs of the people that they are working for. We hope that this issue of *Appropriate Technology,* by highlighting success stories as well as the problems, will stimulate a partnership between these two groups.

❏ Notes

1. Chambers, R., "Rural poverty unperceived ñ problems and remedies." Staff Working Paper No. 400 World Bank. Washington DC, July 1980.
2. Carney, Diana (ed.) "Sustainable rural livelihoods" Report. DFID, London, 1998.
3. IT Transport, "Promoting Intermediate Means of Transport," SSATP Working Paper No. 20. Knowledge, Information and Technology Center, Africa Region, World Bank, Washington DC, 1996.
4. Chingozho, D. "Impact of Intermediate Means of Transport on the allocation of transport and gender relations: Some aspects from an ITDG transport project in Zimbabwe," Case study prepared for the IFRTD's "Balancing the Load: Gender and Transport" programme. IFRTD, London (forthcoming).
5. Matin, Neela, Mahjabeen Chowdhury, Hasina Begum, and Delwara Khanam, "Spatial Mobility and Women's empowerment: Implications for developing rural transport in Bangladesh," Case study prepared for the IFRTD "Balancing the Load: Gender and Transport" programme. IFRTD, London (forthcoming publication).
6. Starkey, P. "Intermediate means of transport: People, paradoxes, and progress." Keynote discussion paper commissioned by the World Bank for expert workshop held June 1999 in Nairoby, Kenya. 52pp, World Bank, Washington DC, forthcoming.
7. UNDP, *Human Development Report.* UNDP New York, 1999.
8. Oram, Colin "When it breaks who can fix it? Issues of ownership and control in engineering for transport." *Forum News* Vol. 6:3 December 1998. IFRTD, London.
9. Gunetilleke, Neranjana "Potential for Commercialising ITSL's projects." Report prepared for the Research and Policy Programme of ITDG Sri Lanka. (unpublished, 1999).
10. Latif, Zainab "The impacts of the use of IMTs on Transport Needs and Access of these by Women in Rural Pakistan." In *Lanka Forum for Rural Transport Development: Meeting Transport Needs with Intermediate Means of Transport Volume II,* 1999. ISBN 9558233013
11. Howe, John *Transport for the Poor or poor transport? A general review of rural transport policy in developing countries with emphasis on low-income areas.* ILO, Geneva, 1997.

12. Iga, Harriet "Impact of bicycle/motorcycle taxi services *(boda-boda)* on women's travel needs in Uganda: A case study of Mpigi District." Case study prepared for IFRTD's "Balancing the Load: Gender and Transport" programme. IFRTD, London (forthcoming publication).
13. Petts, Robert "Maintenance is forever—but how?" *Forum News* Vol. 6:1 June 1998. IFRTD, London.
14. Heggie, Ian "Rural roads: who pays?" Forum News Vol. 6:1 June 1998. IFRTD, London.

FARMLAND PRESERVATION AND SUSTAINABLE AGRICULTURE

GRASSROOTS AND POLICY CONNECTIONS

K. S. Korfmacher

Over the long run, a sustainable food production system requires both a sufficient base of agricultural land and agricultural practices that do not degrade the land. However, current policies and programs for protecting agricultural land are not systematically integrated with those promoting sustainable agriculture. There are various ways that policy-makers, agricultural support organizations, and researchers could better integrate farmland preservation and sustainable agriculture efforts. This paper suggests several approaches for developing such connections including: coordinating local, state, and federal policies, conducting related research, and developing integrated outreach and education programs.

Two issues that are receiving increasing attention from natural resource professionals, farmers, and academics alike are farmland preservation and sustainable agriculture. These two movements have several common goals. They share the objectives of maintaining a healthy agricultural economy, preserving rural communities, and protecting the environment. In addition, both the farmland preservation and sustainable agriculture movements involve multiple stakeholders and entail conflicts between the public good and private property rights. Although these two movements are gaining increased public attention, particularly with respect to farming at the urban-rural interface, both remain outside the mainstream of agricultural policy and practice.

Common sense tells us that a sustainable food system requires both the natural resources necessary for producing food and methods of producing food that do not degrade those resources over time. At least for the foreseeable future, resources for producing food will include agricultural land. However, policies for protecting the agricultural resource base and strategies for promoting sustainable agriculture are not systematically integrated in the U.S. Similarly, nongovernmental efforts to address each of these issues are not closely connected. While progress has been made on both fronts, neither successful farmland preservation nor widespread expansion of environmentally sound agricultural practices alone can achieve a food system that is

From *American Journal of Alternative Agriculture*, Vol. 15, No. 1, by K. S. Korfmacher. Copyright © 2000 by American Journal of Alternative Agriculture.

sustainable for the long run. However, joining the efforts of the two movements could significantly increase the sustainability of our agricultural system.

The current separation between farmland preservation and sustainable agriculture is reflected in the relevant literature. For example, research on sustainable agriculture tends to focus on the development and diffusion of economically viable sustainable practices (Allen, 1993; Hatfield and Karlen, 1993). Although the financial sustainability of farming is acknowledged to be essential to preventing farmland conversion, sustainable agriculture researchers have not explicitly addressed the potential of farmland preservation programs to enhance the economic stability of sustainable farming operations. Similarly, research on farmland preservation has given little attention to the environmental impacts of the farming operations that are being preserved (Daniels and Bowers, 1997). These gaps in the existing literature reflect the lack of connections in practice between policies and organizations supporting sustainable agriculture. This commentary argues that policymakers, agricultural support organizations, and researchers should take steps to integrate farmland preservation and sustainable agriculture efforts and suggests several approaches for developing such connections.

Farmland Preservation

The current separation of farmland preservation and sustainable agriculture efforts can be traced in part to how they have developed historically. That is, farmland preservation efforts have largely been motivated by concerns about the effects of growth on rural landscapes, economies, and communities. In contrast the sustainable agriculture movement has grown primarily from issues of public and environmental health.

Farmland preservation programs respond to several related concerns about land use change. Although few believe that U.S. food supplies are imminently threatened by conversion of prime agricultural lands, some long-term global trend analyses indicate that food supplies may not be able to keep up with the planet's growing population (Faeth, 1997; Gardner, 1996). In addition, the current rate of farmland lost to development in the U.S. is of concern for several reasons. First, the land lost to development is often highly productive land (Daniels and Bowers, 1997). Second, the loss of agricultural capacity near urban centers worries those who predict greater future reliance on local food production due to increased transportation costs, demand for local produce, or climate change. Third, while intensification of agriculture has allowed U.S. farmers to produce more food on roughly the same amount of land over the past four decades, many are skeptical that this trend can continue (Ervin et al., 1998). In addition, most practices necessary for intensification of agriculture historically have resulted in greater environmental damage. These concerns have resulted in efforts over the past 40 years to preserve valuable agricultural lands (Daniels and Bowers, 1997).

While farmland preservation policies have been in place in many states for several decades (Buckland, 1987; Daniels and Bowers, 1997; Klein and Reganold, 1997; Nelson, 1992), recent statistics showing rapid loss of agricultural lands due to urban sprawl have sparked a new wave of public concern about farmland preservation (Daniels, 1999; Krieger, 1999). Efforts to preserve farmland encompass a wide variety of tools and approaches to protect agricultural land from encroaching development. These include land use management tools such as agricultural districts, conservation easements, and cluster zoning provisions (Daniels, 1997). At the state level, differential taxation programs and right-to-farm laws are common (Daniels and Bowers, 1997). In addition, nonprofit groups like the American Farmland Trust promote farmland protection through purchase of development rights and estate planning (American Farmland Trust, 1997; American Farmland Trust, 1999a). Another approach to farmland preservation is increasing the profitability of agriculture through supporting the development of local farmers' markets (Yonkers, 1998).

While farmland preservation has made significant progress towards its goals in some states, the strategies used in the U.S. have been challenged as inefficient and ineffective (McConnell, 1989; Nelson, 1992). Alterman (1997) argues that without a comprehensive policy, farmland preservation efforts will not be significant at a national level in preserving farming economies and communities. In addition, current farmland preservation policies may address some goals, particularly preservation of locally significant landscapes or rural character, more effectively than others, such as protecting environmental values (Kline and Wichelns, 1996; Krieger, 1999). While the public often assumes that preserving farms will protect the environment more than alternative land uses (such as low-density residential development), there is no systematic evidence that this is true. Additionally, few farmland preservation programs control what kinds of agricultural practices and environmental impacts may occur on preserved lands. Thus, while farmland preservation efforts enjoy strong support from opponents of urban sprawl, such programs' overall effectiveness in maintaining viable farm economies, preserving rural communities, and protecting environmental values is uncertain.

Sustainable Agriculture

While the farmland preservation movement is primarily motivated by growth management concerns, interest in sustainable agriculture arises from the relationships among environmental quality, profitability of farming, quality of life issues for farmers, and consumers' food safety and quality concerns (Ervin et al., 1998; Faeth, 1997; Hatfield and Karlen, 1993). Most broadly speaking, sustainable agriculture refers to producing food in a way that does not degrade natural resources or rely on resources that are non-renewable in the long run. As noted by Batie and Taylor (1991), a variety of terms including sustainable agriculture, regenerative agriculture, low-input agriculture, organic agriculture, and eco-agriculture are used to refer to alternatives to conventional, high-input agriculture. Because the focus of this commentary is on encouraging practices that reduce the total environmental costs of agriculture, the term "sustainable agriculture" is used here in its most general sense.

It is impossible to define sustainable agriculture in an absolute way because the term refers to a spectrum of diverse practices that minimize the externalities of farming. These practices are constantly evolving with increasing knowledge and technology (Faeth, 1997). In addition, different people have widely varying conceptions of what is sustainable. For example, while soil conservation practices may be a first step toward reducing the environmental impacts of agriculture, some proponents of sustainable agriculture have as an ultimate goal the elimination of synthetic fertilizers and pesticides. Thus, farms may be more or less sustainable, according to our current understanding and local conditions, but pursuit of sustainability is a continuous process. Given this definition, approaches to sustainable agriculture include, but are not limited to, reducing sediment and nutrient runoff through best management practices, retiring marginal and highly erodible lands, community-supported agriculture (CSA) operations and other efforts to promote local food systems, integrated pest management, and organic farming. Because organic farming is a relatively well-defined and widely understood approach to sustainable agriculture, it is used several times in this commentary as an example of a more sustainable alternative to conventional agriculture. However, the arguments made here apply to a wide range of efforts to improve the sustainability of agriculture.

Reflecting the fact that sustainability is an evolving goal, most programs and organizations that support sustainable agriculture focus on the accumulation and dissemination of knowledge about sustainable practices. The primary federal program supporting sustainable agriculture is the U.S. Department of Agriculture's Sustainable Agriculture Research and Education (SARE) program (Rawson, 1995). Much of the research on sustainable agriculture involves individual farmers or nonprofit organizations that support sustainable farmers (Hassanein and Kloppenburg, 1995; Lockeretz and Anderson, 1990). In addition, many of these organizations work on developing markets for organically grown produce. They also support farmers through information dissemination, financial planning services, and other kinds of technical assistance.

There are many signs that progress has been made toward sustainable agriculture in recent years. For example, retail sales of organic food have increased 20 percent per year since 1989 and this trend is expected to continue (Bourne, 1999). This trend has implications for the profitability of sustainable agriculture. Organic farming may be increasingly attractive relative to conventional agriculture near urban areas where there is a demand for local, sustainably grown produce. In fact, recent studies have shown that certain farms that have adopted more environmentally sound practices have equaled or exceeded the profitability of conventional operations (Northwest Area Foundation, 1994; White et al., 1994). In addition, some practices, such as low-till systems, are believed by some to be better for the environment than conventional practices and also often save farmers money. However, because farmers do not have to pay for the off-site environmental impacts of their practices, the set of widely accepted practices that increase farmers' profits and protect the environment is currently quite small.

Despite promising signs such as the adoption of conservation practices and the increase in organic production, sustainable agricultural systems still account for a very small proportion of U.S. agricultural production. Given the low rate of participation in alternative agriculture efforts, the movement cannot claim to have significantly affected U.S. agricultural communities and their economies.

Common Goals, Common Failures

Although strategies supporting farmland preservation and sustainable agriculture have developed in very different ways, these movements have several common objectives with respect to protecting agricultural economies, rural communities, and the environment. Both issues raise questions of accounting for the full costs and benefits of agriculture, including how to balance present and future environmental, economic, and social needs. In order to incorporate the full social costs and benefits of current agricultural activities into farmers' decision-making, both farmland preservation and sustainable agriculture require consideration of conflicting values. Conflicting values related to agriculture are often tied to trade-offs between private and public property rights. For example, proponents of sustainable agriculture often note that conventional agriculture has not internalized the full social and environmental costs of its impacts, while conventional agricultural interests counter that environmental regulation of farming infringes upon farmers' private property rights (Thompson, 1998). Similarly, those who advocate public funding of farmland preservation programs argue that the public does not currently pay for all of the benefits provided by farmland, such as open space, rural character, and other environmental amenities. Thus, both movements raise controversial questions about property rights and the allocation of natural resources in agricultural areas. Despite such common goals and challenges, programs that promote farmland preservation are not currently integrated with efforts to encourage sustainable agriculture.

Common Solutions

Thus, despite congruences between the objectives of farmland preservation and sustainable agriculture, there have been few attempts to coordinate the two issues, either at a grassroots level or as policy problems. Nonetheless, due to the common characteristics described above, potential exists for joint solutions at the local, state, and federal government levels, as well as by nongovernmental organizations and individual farmers. These solutions include increasing the profitability of agriculture, incorporating sustainability as a criterion in farmland preservation efforts and other land use tools, utilizing market-based regulatory tools, conducting applied research, and developing joint educational programs.

There are several marketing methods that can improve the economic viability of local agriculture. Many of these efforts have the potential to both preserve farms and reduce their environmental impacts. For example, many localities have sponsored farmers' markets, which are particularly important outlets for small organic producers (Yonkers, 1998). In fact, there is significant overlap between programs that benefit small farms and sustainable farms, especially at the urban-rural interface. Organic farms tend to be smaller than conventional farms and often rely on local markets and high-value products. Care must be taken not to equate small farming with sustainable farming, however, since some larger farms may well be more environmentally sustainable than some small farms. Nonetheless, local farm markets often provide new retail opportunities for small farms while reducing the environmental impacts of transporting and storing produce. Value-added products and specialty marketing for sustainably grown products can also enhance the profitability of farming. For example, New York's Watershed Agricultural Program has developed a marketing program for farmers who help protect New York City's water supply through the "crop with the drop" labeling symbol (Ferguson et al., 1998). Although the overall economic impacts of such efforts have not been assessed, they do appear to be promising.

There are several ways that farmland preservation programs could better promote sustainable agriculture through existing tools such as purchase of development rights, agricultural use taxes, and zoning programs. Most purchase-of-development-rights programs involve making decisions about specific areas and farms to target for protection. In the past, these decisions have seldom explicitly considered the environmental impacts of the farms being preserved. Environmental sensitivity of the land could be made an explicit criterion for prioritization; alternately, sustainable practices could be required on parcels preserved with public money. In addition to state or local governmental programs, nonprofit land trusts could also write environmental protection criteria into their conservation easements. Similarly, agricultural-use-taxation programs could be modified to provide for lower tax rates according to the sustainability of the agriculture being practiced. Thus, a certified organic farm might pay only 10% of assessed taxes, a farm with conventional practices but meeting standards for certain best management practices might pay 50%, and a confined animal feeding operation might pay the full tax rate. Finally, stricter environmental protection requirements could be incorporated into zoning measures such as agricultural districts and cluster zoning/planned unit developments.

There are several market-based regulatory tools that could be harnessed to help internalize some of the external costs of agriculture and development. A first step would be to examine existing regulatory programs and eliminate those that favor conventional practices over more sustainable alternatives. As many agro-environmental policy experts have suggested, it is possible to charge farmers for their environmental impacts through programs such as nutrient taxes or trading (Ervin et al., 1998). Either of these measures would encourage more sustainable agricultural practices. On the other side of the equation, new developments should be charged comprehensive impact fees that account for the full costs of the services they require and impacts they cause. In combination, such regulatory tools could change the economics of agriculture so that it is more profitable to continue to farm sustainably in developing areas.

The sustainable agriculture movement has a long history of effective on-farm research. This research should be expanded to encompass farmland preservation issues and to inform both policymakers and individual farmers. Policymakers need to know more about the environmental, economic, and social implications of various kinds of agricultural systems versus new developments. Specifically, a national survey evaluating existing projects and programs that integrate sustainable agriculture and farmland preservation would be helpful to those wishing to promote such connections. In addition, farmers could benefit from continued research on the economic potential of alternative agricultural systems under various conditions. Thus, it may be especially productive to focus on the profitability of sustainable farms (and hence their potential contribution to farmland retention) at the urban/rural interface.

It is important to support educational and outreach programs in conjunction with relevant research. While government programs and existing nonprofit organizations could reorient their strategies to jointly promote farmland preservation and sustainable agriculture, farmers' decisions about agricultural practices and the fate of their lands depend in part on the information available to them and on their judgments about the future. If there are ways that farmers can simultaneously reduce their environmental impacts, increase their profitability, and ensure preservation of their lands, these opportunities should be communicated to a wide range of farmers.

Farmland preservation programs and sustainable agriculture organizations both have well-developed networks that could be used to communicate their common objectives. While neither may be large enough to affect the core of agro-environmental policy, joining forces could enhance their combined impact. For example, many farmland preservation programs have close relationships with state agencies and traditional farm-support organizations. These networks could disseminate information about profitable sustainable-agriculture opportunities to a new audience. Similarly, sustainable agriculture organizations tend to include small farmers who may not be involved in, or aware of, farmland preservation opportunities. Thus, a joint educational program focusing on the connections between sustainable agriculture and farmland preservation might help a wider range of farmers make well-informed choices about how their lands are used. The rising visibility of urban sprawl on the national political agenda bodes well for the potential influence of such joint efforts.

Barriers to Joining Forces

While some of these proposals appear to require only incremental shifts in existing programs, others entail radical changes in current agricultural policy. Predictably, there are many potential barriers to adopting such changes. For example, a key challenge to incorporating sustainability criteria into farmland preservation programs is the difficulty in defining and monitoring sustainable agriculture. That is, in order to tie farmland preservation benefits to sustainable practices on the preserved lands, as suggested above, there must be concrete standards defining sustainable practices. The challenges noted above in defining sustainable agriculture have become an increasingly difficult barrier to policies promoting sustainable agriculture (Youngberg et al., 1993). This challenge is not only technical, but also involves trade-offs between conflicting values. For example, no-till agriculture reduces erosion, improves soil conditions, and minimizes petroleum use; however, it may require more herbicides. Thus, comparing the overall environmental costs of no-till versus conventional production requires a value-based decision about the relative costs of soil loss and carbon dioxide emissions versus herbicide use. Even if such standards were established, making sure they were met would require farmland preservation programs to vastly expand their monitoring and enforcement capabilities. Another challenge would be that such criteria for maintaining sustainable practices might interfere with the flexibility farmers need to adjust to changing technologies and markets.

In addition to the challenges of defining and monitoring sustainable practices, there are several reasons why some proponents of sustainable agriculture and advocates for farmland preservation might not wish to join efforts. For example, proponents of farmland preservation may fear that requiring sustainable practices on preserved parcels may reduce farmers' willingness to participate in farmland preservation programs. Similarly, sustainable agriculture organizations might view association with a farmland preservation program as an additional deterrent to otherwise interested farmers who wish to maximize their flexibility with respect to future use of their lands.

Thus, it is not clear who could best promote the integration of farmland preservation and how. This uncertainty is partly due to the fact that sustainable agriculture and farmland preservation efforts are currently embedded in very different policy subsystems with little overlap. Although there are exceptions, in many states farmland preservation and sustainable agriculture programs are housed in separate agencies or organizations. Farmland preservation programs tend to be staffed by planners in departments of agriculture or administered through county-level task forces concerned with development and land use. Sustainable agriculture programs often involve grassroots organizations of farmers, university researchers, extension agents, and environmentalists. The fact that these two program areas have different professional backgrounds, varied target audiences/clients, and limited experience working together may pose a barrier to joint promotion of the suggested strategies.

Promising Steps

Despite these barriers, there are several signs that increasing numbers of farmers, farming organizations, and policymakers are beginning to grapple with the connections between sustainable agriculture and farmland preservation. Not surprisingly, many of these initiatives are occurring at the urban-rural interface.

There are several reasons why these movements are most salient at the urban-rural interface. As argued above, sustainable farmers may choose to locate in such areas because of the demand for high-quality local produce and reduced transportation and storage costs. This demand is reflected in the growth of local farmers' markets, as discussed above, as well as the emergence of community-supported agriculture (CSA) operations in many urban areas. In addition, for farmers already located in areas where development is encroaching on their farmlands, adopting low-input techniques may be attractive because they are more compatible with socioeconomic changes. For example, some farmers believe that conventional practices such as chemical spraying and intensive feeding operations create more conflict with the new neighbors than do alternatives such as grass-based dairies and organic production systems (Petrucci, 1998). In addition, as land values increase, the economic attractiveness of the high-value-per-acre crops that characterize many alternative agriculture operations increases relative to conventional farming. Thus, there are several economic and social reasons why sustainable practices may be increasingly attractive at the urban/rural interface.

At the same time, farmland preservation efforts are frequently targeted at the urban/rural interface. These areas are often ripe for farmland preservation programs because of the threats of land conversion due to urban sprawl. Farmers in these areas may find their land values soaring as development pressures increase, or that lands they leased in the past are no longer available. These farmers could benefit from programs that aim to preserve farmland through reduced taxes, purchase of development rights, etc. Therefore, both farmland preservation and sustainable agriculture are receiving increasing attention in rapidly urbanizing areas. Despite this trend, links between sustainable agriculture and farmland preservation are emerging in more rural areas as well.

While sustainable agriculture and farmland preservation are not systematically integrated, there are promising individual cases in which the two goals have been successfully linked. The American Farmland Trust simultaneously promotes farmland preservation and sustainable agriculture through many of its programs, including its annual Steward of the Land Award (American Farmland Trust, 1999b). In addition, the Pennsylvania Association for Sustainable Agriculture has supported the American Farmland Trust's efforts to implement sustainable practices on its Cove Mountain Farm (Peterson, 1996; Petrucci, 1998). Farmland preservation programs in Maryland have required all participating farming operations to have a soil and water conservation plan (American Farmland Trust, 1997; Chesapeake Farms for the Future Board, 1998; Pelzman, 1998).

New York City has embarked on an ambitious farmland preservation program to protect its water supply, by encouraging environmentally sound practices on farms in the Catskills watersheds that provide its drinking water (Ferguson, 1998; Ferguson et al., 1998). This project encourages whole farm planning, a promising tool for promoting the common goals of sustainable agriculture and farmland preservation (Minnesota Project, 1996; Watershed Agricultural Council, 1996). Because whole farm planning encourages farmers to develop integrated strategies to manage the profitability, environmental protection, and long-term use of their lands, it encourages farmers to integrate their farming plans and practices with financial and legal methods for protecting their lands from development.

One of the most promising signs for long-term sustainability of our food system is the growing number of creative, forward-thinking farming entrepreneurs who recognize the complementarities between sustainable agriculture and farmland preservation, especially in the urban/rural interface. For example, one farmer from a rapidly developing area of northeastern Ohio converted his high-chemical dairy farm to a diverse organic dairy combined with a small produce farm and vermiculture operation. He is actively involved in the local farmland-preservation task force because, as he says, "I get mad when people say farming is dead! Even though the city is coming in, we can educate the surrounding city and ourselves about the desires and needs of new residents. When we really listen to each other we want basically the same thing. We need new businesses and they need us." Such individual insights about alternative ways of farming profitably while causing fewer negative environmental impacts may make the most important near-term contributions to promoting both sustainable agriculture and farmland preservation. Policymakers and farming organizations should focus on how best to support such innovators.

While efforts to connect sustainable agriculture and farmland preservation appear to be in an embryonic stage, they may represent an emerging dialogue among farmers, nongovernmental organizations, natural resource agencies, and the public about how to promote an environmentally, socially, and economically sustainable food system. As Ervin et al. (1998) argue, given strong public support for protecting the environment and the likelihood of increased demand for agricultural products in the near future, it is time to envision a new policy approach to reducing the diverse environmental impacts of agriculture. This must be done in a way that enhances the profitability of agricultural operations while accounting for their full environmental and social costs. In order to achieve the goal of an environmentally sustainable food system, new agricultural policies as well as nongovernmental efforts should promote both farmland preservation and sustainable agriculture in integrated ways.

❏ References

Allen, P. (ed.). 1993. Food for the Future: Conditions and Contradictions of Sustainability. John Wiley & Sons, New York.

Alterman, R. 1997. The challenge of farmland preservation: Lessons from a six-nation comparison. J. Amer. Planning Assoc. 63:220–243.

American Farmland Trust. 1997. Saving American Farmland: What Works. Northampton, MA.

American Farmland Trust. 1999a. Home page. Web site http://www. farmland.org (accessed Oct. 1999).

American Farmland Trust. 1999b. $10,000 Steward of the Land Award. Web site http://www.farmland.org/ files/steward/steward.htm (accessed Oct. 1999).

Batie, S.S., and D.B. Taylor. 1991. Assessing the character of agricultural production systems: Issues and implications. Amer. J. Alternative Agric. 6:184–187.

Bourne, J. 1999. The organic revolution. Audubon (March):64–70.

Buckland, J.G. 1987. The history and use of purchase of development rights in the United States. Landscape and Urban Planning 14:237–252.

Chesapeake Farms for the Future Board. 1998. Farms for the Future: A Strategic Approach to Saving Maryland's Farmland and Rural Resources. American Farmland Trust, Washington, DC.

Daniels, T.L. 1997. Where does cluster zoning fit in farmland protection? J. Amer. Planning Assoc. 63:129–137.

Daniels, T.L. 1999. When City and Country Collide: Managing Growth in the Metropolitan Fringe. Island Press, Washington, DC.

Daniels, T.L., and D. Bowers. 1997. Holding Our Ground: Protecting America's Farms and Farmlands. Island Press, Washington, DC.

Ervin, D.E., C.F. Runge, E.A. Graffy, W.E. Anthony, S.S. Batie, P. Faeth, T. Penny, and T. Warman. 1998. Agriculture and environment: A new strategic vision. Environment 40(6):8–15, 35–40.

Faeth, P. 1997. Sustainability and U.S. agriculture: Problems, progress, and prospects. In R.C. Dower, D. Ditz, P. Faeth, H. Johnson, K. Kozloff, and J.J. MacKenzie (eds.). Frontiers of Sustainability: Environmentally Sound Agriculture, Forestry, Transportation, and Power Production. Island Press, Washington, DC.

Ferguson, K. 1998. Upstream New York: Preserving upstate farms to protect New York City's water supply. American Farmland 19(3):14–17.

Ferguson, K., J.P. Cosgrove, and T. Ptacek. 1998. Call to Action: Farmland Protection Success Stories in the Empire State. American Farmland Trust, Saratoga Springs, NY.

Gardner, G. 1996. Shrinking Fields: Cropland Loss in a World of Eight Billion. Worldwatch Paper 131. Worldwatch Institute, Washington, DC.

Hassanein, N., and J.R. Kloppenburg, Jr. 1995. Where the grass grows again: Knowledge exchange in the sustainable agriculture movement. Rural Soc. 60:721–740.

Hatfield J.L., and D.L. Karlen (eds.). 1993. Sustainable Agriculture Systems. Lewis Publishers, Boca Raton, FL.

Klein, L.R., and J.P. Reganold. 1997. Agricultural changes and farmland protection in western Washington. J. Soil Water Conserv. 52:6–12.

Kline, J., and D. Wichelns. 1996. Public preferences regarding the goals of farmland preservation programs. Land Econ. 72:538–549.

Krieger, D.J. 1999. Saving Open Spaces: Public Support for Farmland Protection. Working Paper CAE/WP99–1. American Farmland Trust Center for Agriculture in the Environment, DeKalb, IL. Available at Web site http://farm.fic.niu.edu/cae/wp/99-1/wp99-1.html (accessed Oct. 1999).

Lockeretz, W., and M.D. Anderson. 1990. Farmers' role in sustainable agriculture research. Amer. J. Alternative Agric. 5:178–183.

McConnell, K.E. 1989. The optimal quantity of land in agriculture. Northeastern J. Agric. Res. Econ. 18:63–72.

Minnesota Project. 1996. What is comprehensive farm planning? Whole Farm Planner 1(3):6. St. Paul, MN. Available at Web site http://www.misa.umn.edu/wfpv1n3.html (accessed Oct. 1999).

Nelson, A.C. 1992. Preserving prime farmland in the face of urbanization: Lessons from Oregon. J. Amer. Planning Assoc. 58:467–488.

Northwest Area Foundation. 1994. A Better Row to Hoe: The Economic, Environmental, and Social Impact of Sustainable Agriculture. St. Paul, MN.

Pelzman, H. 1998. Mapping the future. American Farmland 19(2):16–22.

Peterson, J.D. 1996. The green, green grass of home. American Farmland (Summer):4–6.

Petrucci, B. 1998. Greener pastures. American Farmland 19(2):12–15.

Rawson, J.M. 1995. Sustainable Agriculture. Report no. 95–1062. Library of Congress, Congressional Research Service, Washington, DC. Available at Committee for the National Institute for the Environment Web site http//www.cnie.org/nle/ag-14.html (accessed Oct. 1999).

Thompson, E., Jr. 1998. Sharing the Responsibility: What Agricultural Landowners Think About Property Rights, Government Regulation and the Environment. American Farmland Trust, Washington, DC.

Watershed Agricultural Council. 1996. Whole Farm Planning. New York City Dept. of Environmental Protection, Watershed Agricultural Program, Walton, NY.

White, D.C., J.B. Braden, and R.H. Hornbaker. 1994. Economics of sustainable agriculture. In J.L. Hatfield and D.L. Karlen (eds.). Sustainable Agriculture Systems. Lewis Publishers, Boca Raton, FL.

Yonkers, A.H. 1998. Selling direct. American Farmland 19(1):20–21.

Youngberg, G., N. Schaller, and K. Merrigan. 1993. The sustainable agriculture policy agenda in the United States: Politics and prospects. In P. Allen (ed.). Food for the Future: Conditions and Contradictions of Sustainability. John Wiley & Sons, New York. p. 295–318.

APPENDIX

Possible Assignments for Critical Thinking and Reading

VIEWPOINT

Select an article and answer the following questions. When you have finished answering the questions, write a short essay expressing your personal feeling about the subject, making sure to give specifics from the article to support your opinion.

1. What are the major facts relating to the issue discussed in the article?

2. How do you think the issue discussed in this article impacts agriculture?

3. What do you think the future significance of this issue will be for your major?

4. What is your opinion of this issue?

5. List supporting arguments and/or details for your position?

6. What will be your introduction for this essay?

7. What will be your conclusion for this essay?

Now write the essay.

READING AN ARTICLE CRITICALLY

Select an article and read through it carefully while looking for the answers to some of the questions below. Then re-read the article finding the answers to the following questions. You can answer the questions in paragraph form or by numbers.

1. What is the main point of the article? Your answer must include
    a. A subject—what or who is the article about?
    b. A point about the subject—how is the subject being affected?

2. What is the author's purpose for writing this article (to explain, describe, argue, entertain, persuade, inform)?

3. What is the author's tone (angry, critical, neutral)?

4. Give three facts from the article.

5. Give three examples of opinions from the article (if they are present).

6. What source(s) does the author use to support his/her opinions (if any)?

7. Does the author's reasoning behind his/her opinions seem rational or weak? Explain.

8. What questions does this article bring to light about the subject of the article?

9. In a short paragraph (about 5 sentences) express your view regarding the subject of the article, making sure to give specific support from the article to back your view.

GUIDING YOUR READING

Read an article and answer the following questions.

1. What is the key issue of the article?

2. Who is the intended audience?

3. Does the article make an argument for or against something or does it just state facts? What are the arguments or the facts?

4. What is the big picture presented in the article?

5. What material can you filter out which is not totally relevant to the point being made in the article?

6. How logical is the article? What makes it logical or confusing?

7. Is there a strong conclusion?

WHAT IS ANALYSIS IN WRITING?

To analyze something is to ask what that something *means*.

Analysis means suspending judgment . . . not giving your opinion at first.

Analysis means making *explicit (state openly)* what the article says is *implicit (suggested but not stated openly)*.

Then you ask the question "What do I *infer* from the reading?"

Exercise:

Read an article.

1. List what is suggested but not overtly stated by the article?

2. List what is explicit (openly acknowledged) to you?

3. What do you *infer* from the article?

4. What is it about this article that makes it special, and why did the author choose these ideas he or she wrote about?

5. What words are repeated or similar to each other in this article?

This is one method of analyzing an article to find its meaning.

CRITICAL THINKING PLAN

Read an article and analyze the major points using the following plan.

FACTS   (Problem/facts)

PATTERNS   (What is seen as the big picture which the
                               article addresses?)

HYPOTHETICAL   (What questions do you, as the

THINKING   reader, have about the topic?)

IMPLICATIONS   (What implications can you see about the
                               topic?)

FILTER   (Filter out what isn't necessary)

SOLUTIONS/CONCLUSIONS   (What does the author conclude? Do you
                               agree? Why or why not—be specific.)

ELEMENTS OF THOUGHT

Read an article and analyze the major points using the following plan:

POINT OF VIEW OF THE AUTHOR

PURPOSE OF THE ARTICLE

ISSUE OR PROBLEM

CONCEPT BEING EXPRESSED

ASSUMPTIONS (INFERENCES) YOU MADE

CONSEQUENCES/OUTCOMES OF THE AUTHOR

READINGS IN
AGRICULTURE

Answer the following questions . . . in detail:

1. WHAT did the author say about the topic? (You must make sure and tell WHAT the article is about.)

2. HOW did the author come to the conclusion expressed in the article? (You should mention facts which are explicit and opinions which may be inferred or implied.)

3. WHY is (are) the conclusion(s) in the article important to the agriculture industry/business? (This section may be your opinion and can be inferred from the facts and conclusions in the article.)

ASSIGNMENT FOR READINGS IN AGRICULTURE

Find the main point of the article you have read. Make at least 10 observations from the article about that main point.

Your paper should look like this:

Name

Article Title
Author

Main point:

   1.

   2.

   3.

   4.

   5.

   6.

   7.

   8.

   9.

  10.

READINGS IN AGRICULTURE

Choose 3 articles that talk about a problem (maybe a solution). Please try to choose articles which have not been reported on in class before.

Do the following with each article:

1. State what the article is about.

2. What exactly is the problem being discussed in the article?

3. What choices are there for solutions? (You may need to infer what you believe might be some solutions.)

4. What are the advantages and disadvantages of each solution you mentioned in 3? You can use list or column format.

5. What seems to be the best choice and why?

Writing a Research Paper

Do Some Groundwork

Identify the basic requirements for the paper. These might include the following:

A. Topic restrictions, if any

B. Paper length

C. Required number of sources

D. Types of sources required

E. Format requirements

F. Material to be turned in with the paper

G. Due date, or due dates if you have to turn in parts before the final paper

Prepare a project breakdown. This should include due dates to help pace yourself through this type of lengthy assignment.

Choose your topic carefully.

Gather your material/Including interviews

1. Schedule some reading time.

2. Take notes on the reading material.

3. Plan interviews.

4. Synthesize material.

Make an Outline

Write the Paper

1. Write a rough draft/write another rough draft/write another rough draft.

2. Leave the rough draft for a time.

3. Write a final copy

4. Use *in-text notes* and a *Works Cited* page.

5. Proofread the paper before submitting

6. Always make and keep a photo or computer copy of the paper before turning it in (you might even save it on two different disks).

SUMMARY WRITING

* Name the source of the material being summarized—for example, title, author, name of publication and date of publication.

* Tell about the subject of the article and what the author's main point(s) is (are).

* Elaborate on the author's main point(s).

If you are asked to give your opinion, after you summarize the main point(s), use each point and give your opinion. Make sure to be specific, using examples about which the author speaks, for your opinion.

READINGS IN AGRICULTURE

Choose an article and use specific information to fill in the sections below.

The author says:

The teacher says:

The class members say:

I say:

Synthesis: (This section integrates what has been said).